High-Yield™

Biostatistics, Epidemiology, &
Public Health

FOURTH EDITION

High-Yield™
Biostatistics, Epidemiology, & Public Health

FOURTH EDITION

Anthony N. Glaser, MD, PhD
Clinical Assistant Professor of Family Medicine
Department of Family Medicine
Medical University of South Carolina
Charleston, South Carolina

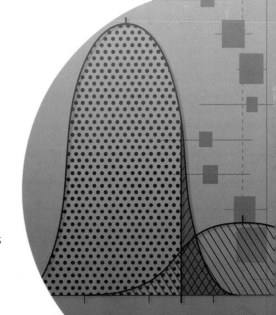

Wolters Kluwer | Lippincott Williams & Wilkins
Health

Philadelphia • Baltimore • New York • London
Buenos Aires • Hong Kong • Sydney • Tokyo

Acquisitions Editor: Susan Rhyner
Product Manager: Catherine Noonan
Marketing Manager: Joy Fisher-Williams
Vendor Manager: Bridgett Dougherty
Manufacturing Manager: Margie Orzech
Design Coordinator: Teresa Mallon
Compositor: S4Carlisle Publishing Services

Fourth Edition

Library of Congress Cataloging-in-Publication Data
Glaser, Anthony N.
    [High-yield biostatistics]
    High-yield biostatistics, epidemiology, and public health / Anthony N. Glaser, MD, PhD, clinical assistant professor, Medical University of South Carolina. — 4th edition.
        pages cm
    Earlier title: High-yield biostatistics.
    Includes bibliographical references and index.
    ISBN 978-1-4511-3017-1
    1. Medical statistics.   2. Biometry.   I. Title.
    R853.S7G56 2014
    570.1'5195—dc23
                                                                    2012039198

DISCLAIMER

Care has been taken to confirm the accuracy of the information present and to describe generally accepted practices. However, the authors, editors, and publisher are not responsible for errors or omissions or for any consequences from application of the information in this book and make no warranty, expressed or implied, with respect to the currency, completeness, or accuracy of the contents of the publication. Application of this information in a particular situation remains the professional responsibility of the practitioner; the clinical treatments described and recommended may not be considered absolute and universal recommendations.

The authors, editors, and publisher have exerted every effort to ensure that drug selection and dosage set forth in this text are in accordance with the current recommendations and practice at the time of publication. However, in view of ongoing research, changes in government regulations, and the constant flow of information relating to drug therapy and drug reactions, the reader is urged to check the package insert for each drug for any change in indications and dosage and for added warnings and precautions. This is particularly important when the recommended agent is a new or infrequently employed drug.

Some drugs and medical devices presented in this publication have Food and Drug Administration (FDA) clearance for limited use in restricted research settings. It is the responsibility of the health care provider to ascertain the FDA status of each drug or device planned for use in their clinical practice.

To purchase additional copies of this book, call our customer service department at (800) 638-3030 or fax orders to (301) 223-2320. International customers should call (301) 223-2300.

Visit Lippincott Williams & Wilkins on the Internet: http://www.lww.com. Lippincott Williams & Wilkins customer service representatives are available from 8:30 am to 6:00 pm, EST.

RRS1212

9 8 7 6 5 4 3 2 1

To my wife, Marlene

Contents

WITHDRAWN

Statistical Symbols ....inside front cover
Preface ix

1 Descriptive Statistics 1

Populations, Samples, and Elements 1
Probability.. 1
Types of Data.. 2
Frequency Distributions 3
Measures of Central Tendency... 8
Measures of Variability... 9
Z Scores... 12

2 Inferential Statistics 15

Statistics and Parameters 15
Estimating the Mean of a Population 19
t Scores 21

3 Hypothesis Testing 24

Steps of Hypothesis Testing 24
Z-Tests.. 28
The Meaning of Statistical Significance 28
Type I and Type II Errors 28
Power of Statistical Tests... 29
Directional Hypotheses... 31
Testing for Differences between Groups... 32
Post Hoc Testing and Subgroup Analyses... 33
Nonparametric and Distribution-Free Tests 34

4 Correlational and Predictive Techniques 36

Correlation.. 36
Regression 38
Survival Analysis... 40
Choosing an Appropriate Inferential or Correlational Technique 43

5 Asking Clinical Questions: Research Methods 45

Simple Random Samples.. 46

Stratified Random Samples 46
Cluster Samples 46
Systematic Samples 46
Experimental Studies 46
Research Ethics and Safety 51
Nonexperimental Studies 53

6 Answering Clinical Questions I: Searching for and Assessing the Evidence 59

Hierarchy of Evidence 60
Systematic Reviews 60

7 Answering Clinical Questions II: Statistics in Medical Decision Making 68

Validity 68
Reliability 69
Reference Values 69
Sensitivity and Specificity 70
Receiver Operating Characteristic Curves 74
Predictive Values 75
Likelihood Ratios 77
Prediction Rules 80
Decision Analysis 81

8 Epidemiology and Population Health 86

Epidemiology and Overall Health 86
Measures of Life Expectancy 88
Measures of Disease Frequency 88
Measurement of Risk 92

9 Ultra-High-Yield Review 101

References 105
Index 107

Preface

This book aims to fill the need for a short, down-to-earth, high-yield survey of biostatistics, and judging by the demand for a fourth edition, it seems to have succeeded so far.

One big change in this edition: in anticipation of an expected major expansion of the material to be included in the USMLE Content Outline, with the inclusion of Epidemiology and Population Health, this book covers much more material. The USMLE (US Medical Licensing Examination) is also focusing more and more on material that will be relevant to the practicing physician, who needs to be an intelligent and critical reader of the vast amount of medical information that appears daily, not only in the professional literature but also in pharmaceutical advertising, news media, and websites, and are often brought in by patients bearing printouts and reports of TV programs they have seen. USMLE is taking heed of these changes, which can only be for the better.

This book aims to cover the complete range of biostatistics, epidemiology, and population health material that can be expected to appear in USMLE Step 1, without going beyond that range. For a student who is just reviewing the subject, the mnemonics, the items marked as high-yield, and the ultra-high-yield review will allow valuable points to be picked up in an area of USMLE that is often neglected.

But this book is not just a set of notes to be memorized for an exam. It also provides explanations and (I hope) memorable examples so that the many medical students who are confused or turned off by the excessive detail and mathematics of many statistics courses and textbooks can get a good understanding of a subject that is essential to the effective practice of medicine.

Most medical students are not destined to become producers of research (and those that do will usually call on professional statisticians for assistance)—but all medical decisions, from the simplest to the most complex, are made in the light of knowledge that has grown out of research. Whether we advise a patient to stop smoking, to take an antibiotic, or to undergo surgery, our advice must be made on the basis of some kind of evidence that this course of action will be of benefit to the patient. How this evidence was obtained and disseminated, and how we understand it, is therefore critical; there is perhaps no other area in USMLE Step 1 from which knowledge will be used every day by every physician, no matter what specialty they are in, and no matter what setting they are practicing in.

I have appreciated the comments and suggestions about the first three editions that I have received from readers, both students and faculty, at medical schools throughout the United States and beyond. If you have any ideas for changes or improvements, or if you find a biostatistics question on USMLE Step 1 that you feel this book did not equip you to answer, please drop me a line.

Anthony N. Glaser, MD, PhD
tonyglaser@mindspring.com

Chapter 1

Descriptive Statistics

Statistical methods fall into two broad areas: **descriptive statistics** and **inferential statistics**.

- **Descriptive statistics** merely describe, organize, or summarize data; they refer only to the actual data available. Examples include the mean blood pressure of a group of patients and the success rate of a surgical procedure.
- **Inferential statistics** involve making inferences that go beyond the actual data. They usually involve inductive reasoning (i.e., generalizing to a population after having observed only a sample). Examples include the mean blood pressure of all Americans and the expected success rate of a surgical procedure in patients who have not yet undergone the operation.

Populations, Samples, and Elements

A **population** is the universe about which an investigator wishes to draw conclusions; it need not consist of people, but may be a population of measurements. Strictly speaking, if an investigator wants to draw conclusions about the blood pressure of Americans, the population consists of the blood pressure measurements, not the Americans themselves.

A **sample** is a subset of the population—the part that is actually being observed or studied. Researchers can only rarely study whole populations, so inferential statistics are almost always needed to draw conclusions about a population when only a sample has actually been studied.

A single observation—such as one person's blood pressure—is an **element**, denoted by **X**. The number of elements in a population is denoted by **N**, and the number of elements in a sample by **n**. A population therefore consists of all the elements from X_1 to X_N, and a sample consists of n of these N elements.

Probability

The **probability** of an event is denoted by **p**. Probabilities are usually expressed as decimal fractions, not as percentages, and must lie between zero (zero probability) and one (absolute certainty). The probability of an event cannot be negative. The probability of an event can also be expressed as a ratio of the number of likely outcomes to the number of possible outcomes.

> For example, if a fair coin were tossed an infinite number of times, heads would appear on 50% of the tosses; therefore, the probability of heads, or p (heads), is .50. If a random sample of 10 people were drawn an infinite number of times from a population of 100 people, each person would be included in the sample 10% of the time; therefore, p (being included in any one sample) is .10.

The probability of an event *not* occurring is equal to one minus the probability that it will occur; this is denoted by **q**. In the above example, the probability of any one person *not* being included in any one sample (q) is therefore $1 - p = 1 - .10 = .90$.

> The USMLE requires familiarity with the three main methods of calculating probabilities: the addition rule, the multiplication rule, and the binomial distribution.

Addition rule

*The **addition rule** of probability states that the probability of any one of several particular events occurring is equal to the sum of their individual probabilities, provided the events are mutually exclusive (i.e., they cannot both happen).*

Because the probability of picking a heart card from a deck of cards is .25, and the probability of picking a diamond card is also .25, this rule states that the probability of picking a card that is either a heart or a diamond is .25 + .25 = .50. Because no card can be both a heart and a diamond, these events meet the requirement of mutual exclusiveness.

Multiplication rule

*The **multiplication rule** of probability states that the probability of two or more statistically independent events all occurring is equal to the product of their individual probabilities.*

If the lifetime probability of a person developing cancer is .25, and the lifetime probability of developing schizophrenia is .01, the lifetime probability that a person might have both cancer and schizophrenia is .25 × .01 = .0025, provided that the two illnesses are independent—in other words, that having one illness neither increases nor decreases the risk of having the other.

BINOMIAL DISTRIBUTION

The probability that a *specific combination of mutually exclusive independent events* will occur can be determined by the use of the **binomial distribution**. A binomial distribution is one in which there are only two possibilities, such as yes/no, male/female, and healthy/sick. If an experiment has exactly two possible outcomes (one of which is generally termed *success*), the binomial distribution gives the probability of obtaining an exact number of successes in a series of independent trials.

A typical medical use of the binomial distribution is in genetic counseling. Inheritance of a disorder such as Tay-Sachs disease follows a binomial distribution: there are two possible events (inheriting the disease or not inheriting it) that are mutually exclusive (one person cannot both have and not have the disease), and the possibilities are independent (if one child in a family inherits the disorder, this does not affect the chance of another child inheriting it).

> A physician could therefore use the binomial distribution to inform a couple who are carriers of the disease how probable it is that some specific combination of events might occur—such as the probability that if they are to have two children, neither will inherit the disease.
>
> The formula for the binomial distribution does not need to be learned or used for the purposes of the USMLE.

Types of Data

The choice of an appropriate statistical technique depends on the type of data in question. Data will always form one of four **scales of measurement: nominal, ordinal, interval,** or **ratio**. The mnemonic "NOIR" can be used to remember these scales in order. Data may also be characterized as discrete or continuous.

- **Nominal** scale data are divided into qualitative categories or groups, such as male/female, black/white, urban/suburban/rural, and red/green. There is no implication of order or ratio. Nominal data that fall into only two groups are called dichotomous data.
- **Ordinal** scale data can be placed in a meaningful order (e.g., students may be ranked 1st/2nd/3rd in their class). However, there is no information about the size of the interval—no conclusion can be drawn about whether the difference between the first and second students is the same as the difference between the second and third.
- **Interval** scale data are like ordinal data, in that they can be placed in a meaningful order. In addition, they have meaningful intervals between items, which are usually measured quantities. For example, on the Celsius scale, the difference between 100° and 90° is the same as the difference between 50° and 40°. However, because interval scales do not have an absolute zero, ratios of scores are not meaningful: 100°C is not twice as hot as 50°C because 0°C does not indicate a complete absence of heat.
- **Ratio** scale data have the same properties as interval scale data; however, because there is an absolute zero, meaningful ratios do exist. Most biomedical variables form a ratio scale: weight in grams or pounds, time in seconds or days, blood pressure in millimeters of mercury, and pulse rate in beats per minute are all ratio scale data. The only ratio scale of temperature is the kelvin scale, in which zero indicates an absolute absence of heat, just as a zero pulse rate indicates an absolute lack of heartbeat. Therefore, it is correct to say that a pulse rate of 120 beats/min is twice as fast as a pulse rate of 60 beats/min, or that 300K is twice as hot as 150K.
- **Discrete** variables can take only certain values and none in between. For example, the number of patients in a hospital census may be 178 or 179, but it cannot be in between these two; the number of syringes used in a clinic on any given day may increase or decrease only by units of one.
- **Continuous** variables may take any value (typically between certain limits). Most biomedical variables are continuous (e.g., a patient's weight, height, age, and blood pressure). However, the process of measuring or reporting continuous variables will reduce them to a discrete variable; blood pressure may be reported to the nearest whole millimeter of mercury, weight to the nearest pound, and age to the nearest year.

Frequency Distributions

A set of unorganized data is difficult to digest and understand. Consider a study of the serum cholesterol levels of a sample of 200 men: a list of the 200 measurements would be of little value in itself. A simple first way of organizing the data is to list all the possible values between the highest and the lowest in order, recording the frequency (f) with which each score occurs. This forms a **frequency distribution**. If the highest serum cholesterol level were 260 mg/dL, and the lowest were 161 mg/dL, the frequency distribution might be as shown in Table 1-1.

GROUPED FREQUENCY DISTRIBUTIONS

Table 1-1 is unwieldy; the data can be made more manageable by creating a **grouped frequency distribution**, shown in Table 1-2. Individual scores are grouped (between 7 and 20 groups are usually appropriate). Each group of scores encompasses an **equal class interval**. In this example, there are 10 groups with a class interval of 10 (161 to 170, 171 to 180, and so on).

RELATIVE FREQUENCY DISTRIBUTIONS

As Table 1-2 shows, a grouped frequency distribution can be transformed into a **relative frequency distribution**, which shows the *percentage* of all the elements that fall within each class interval. The relative frequency of elements in any given class interval is found by dividing f, the frequency (or number of elements) in that class interval, by n (the sample size, which in this case is 200).

TABLE 1-1		FREQUENCY DISTRIBUTION OF SERUM CHOLESTEROL LEVELS IN 200 MEN							
Score	f	Score	f	Score	f	Score	f	Score	f
260	1	240	2	220	4	200	3	180	0
259	0	239	1	219	2	199	0	179	2
258	1	238	2	218	1	198	1	178	1
257	0	237	0	217	3	197	3	177	0
256	0	236	3	216	4	196	2	176	0
255	0	235	1	215	5	195	0	175	0
254	1	234	2	214	3	194	3	174	1
253	0	233	2	213	4	193	1	173	0
252	1	232	4	212	6	192	0	172	0
251	1	231	2	211	5	191	2	171	1
250	0	230	3	210	8	190	2	170	1
249	2	229	1	209	9	189	1	169	1
248	1	228	0	208	1	188	2	168	0
247	1	227	2	207	9	187	1	167	0
246	0	226	3	206	8	186	0	166	0
245	1	225	3	205	6	185	2	165	1
244	2	224	2	204	8	184	1	164	0
243	3	223	1	203	4	183	1	163	0
242	2	222	2	202	5	182	1	162	0
241	1	221	1	201	4	181	1	161	1

TABLE 1-2	GROUPED, RELATIVE, AND CUMULATIVE FREQUENCY DISTRIBUTIONS OF SERUM CHOLESTEROL LEVELS IN 200 MEN		
Interval	Frequency f	Relative f	Cumulative f
251–260	5	2.5	100.0
241–250	13	6.5	97.5
231–240	19	9.5	91.0
221–230	18	9.0	81.5
211–220	38	19.0	72.5
201–210	72	36.0	53.5
191–200	14	7.0	17.5
181–190	12	6.0	10.5
171–180	5	2.5	4.5
161–170	4	2.0	2.0

By multiplying the result by 100, it is converted into a percentage. Thus, this distribution shows, for example, that 19% of this sample had serum cholesterol levels between 211 and 220 mg/dL.

CUMULATIVE FREQUENCY DISTRIBUTIONS

Table 1-2 also shows a **cumulative frequency distribution**. This is also expressed as a percentage; it shows the percentage of elements lying *within and below* each class interval. Although a group may be called the 211–220 group, this group actually includes the range of scores that lie from 210.5 up to and including 220.5—so these figures are the **exact upper** and **lower limits** of the group.

The relative frequency column shows that 2% of the distribution lies in the 161–170 group and 2.5% lies in the 171–180 group; therefore, a total of 4.5% of the distribution lies at or below a score of 180.5, as shown by the cumulative frequency column in Table 1-2. A further 6% of the distribution lies in the 181–190 group; therefore, a total of $(2 + 2.5 + 6) = 10.5\%$ lies at or below a score of 190.5. A man with a serum cholesterol level of 190 mg/dL can be told that roughly 10% of this sample had lower levels than his and that approximately 90% had scores above his. The cumulative frequency of the highest group (251–260) must be 100, showing that 100% of the distribution lies at or below a score of 260.5.

GRAPHICAL PRESENTATIONS OF FREQUENCY DISTRIBUTIONS

Frequency distributions are often presented as graphs, most commonly as **histograms**. Figure 1-1 is a histogram of the grouped frequency distribution shown in Table 1-2; the **abscissa** (X or horizontal axis) shows the grouped scores, and the **ordinate** (Y or vertical axis) shows the frequencies.

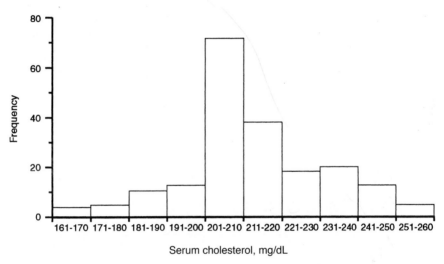

● **Figure 1-1** Histogram of grouped frequency distribution of serum cholesterol levels in 200 men.

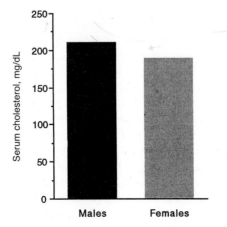

● **Figure 1-2** Bar graph of mean serum cholesterol levels in 100 men and 100 women.

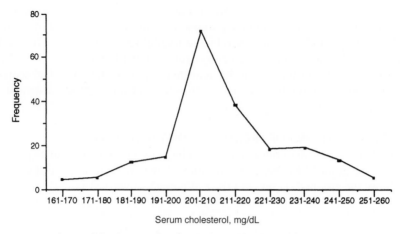

● **Figure 1-3** Frequency polygon of distribution of serum cholesterol levels in 200 men.

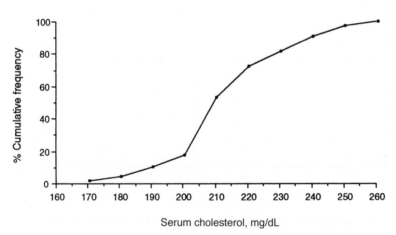

● **Figure 1-4** Cumulative frequency distribution of serum cholesterol levels in 200 men.

To display nominal scale data, a **bar graph** is typically used. For example, if a group of 100 men had a mean serum cholesterol value of 212 mg/dL and a group of 100 women had a mean value of 185 mg/dL, the means of these two groups could be presented as a bar graph, as shown in Figure 1-2.

Bar graphs are identical to frequency histograms, except that each rectangle on the graph is clearly separated from the others by a space, showing that the data form discrete categories (such as male and female) rather than continuous groups.

For ratio or interval scale data, a frequency distribution may be drawn as a **frequency polygon**, in which the midpoints of each class interval are joined by straight lines, as shown in Figure 1-3.

A cumulative frequency distribution can also be presented graphically as a polygon, as shown in Figure 1-4. Cumulative frequency polygons typically form a characteristic S-shaped curve known as an **ogive**, which the curve in Figure 1-4 approximates.

CENTILES AND OTHER QUANTILES

The cumulative frequency polygon and the cumulative frequency distribution both illustrate the concept of **centile** (or **percentile**) **rank**, which states the percentage of observations that fall below

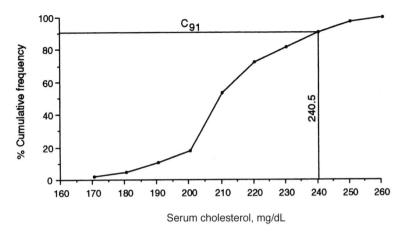

● **Figure 1-5** Cumulative frequency distribution of serum cholesterol levels in 200 men, showing location of 91st centile.

any particular score. In the case of a grouped frequency distribution, such as the one in Table 1-2, centile ranks state the percentage of observations that fall within or below any given class interval. Centile ranks provide a way of giving information about one individual score in relation to all the other scores in a distribution.

For example, the cumulative frequency column of Table 1-2 shows that 91% of the observations fall below 240.5 mg/dL, which therefore represents the 91st centile (which can be written as C_{91}), as shown in Figure 1-5. A man with a serum cholesterol level of 240.5 mg/dL lies at the 91st centile—about 9% of the scores in the sample are higher than his.

Centile ranks are widely used in reporting scores on educational tests. They are one member of a family of values called **quantiles**, which divide distributions into a number of equal parts. Centiles divide a distribution into 100 equal parts. Other quantiles include **quartiles**, which divide the data into 4 parts, **quintiles**, which divide the data into 5 parts, and **deciles**, which divide a distribution into 10 parts.

THE NORMAL DISTRIBUTION

Frequency polygons may take many different shapes, but many naturally occurring phenomena are approximately distributed according to the symmetrical, bell-shaped **normal** or **Gaussian distribution**, as shown in Figure 1-6.

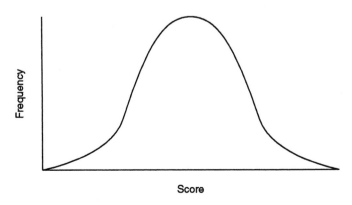

● **Figure 1-6** The normal or Gaussian distribution.

SKEWED, J-SHAPED, AND BIMODAL DISTRIBUTIONS

Figure 1-7 shows some other frequency distributions. Asymmetric frequency distributions are called **skewed** distributions. **Positively** (or **right**) **skewed** distributions and **negatively** (or **left**) **skewed** distributions can be identified by the location of the *tail* of the curve (not by the location of the hump—a common error). Positively skewed distributions have a relatively large number of low scores and a small number of very high scores; negatively skewed distributions have a relatively large number of high scores and a small number of low scores.

Figure 1-7 also shows a **J-shaped** distribution and a **bimodal** distribution. Bimodal distributions are sometimes a combination of two underlying normal distributions, such as the heights of a large number of men and women—each gender forms its own normal distribution around a different midpoint.

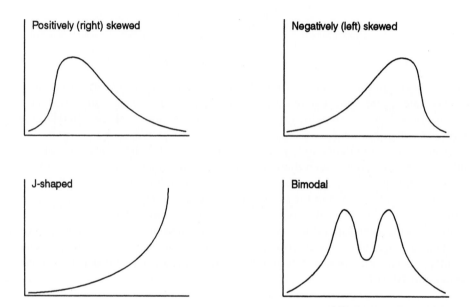

● **Figure 1-7** Examples of nonnormal frequency distributions.

Measures of Central Tendency

An entire distribution can be characterized by one typical measure that represents all the observations—**measures of central tendency**. These measures include the **mode**, the **median**, and the **mean**.

Mode

The mode is the observed value that occurs with the greatest frequency. It is found by simple inspection of the frequency distribution (it is easy to see on a frequency polygon as the highest point on the curve). If two scores both occur with the greatest frequency, the distribution is bimodal; *if more than two scores occur with the greatest frequency, the distribution is* multimodal. *The mode is sometimes symbolized by* **Mo**. *The mode is totally uninfluenced by small numbers of extreme scores in a distribution.*

Median

The median is the figure that divides the frequency distribution in half when all the scores are listed in order. When a distribution has an odd number of elements, the median is therefore the middle one; when it has an even number of elements, the median lies halfway between the two middle scores (i.e., it is the average or mean of the two middle scores).

*For example, in a distribution consisting of the elements 6, 9, 15, 17, 24, the median would be 15. If the distribution were 6, 9, 15, 17, 24, 29, the median would be 16 (the average of 15 and 17). The median responds only to the number of scores above it and below it, not to their actual values. If the above distribution were 6, 9, 15, 17, 24, 500 (rather than 29), the median would still be 16— so the median is insensitive to small numbers of extreme scores in a distribution; therefore, it is a very useful measure of central tendency for highly skewed distributions. The median is sometimes symbolized by **Mdn**. It is the same as the 50th centile (C_{50}).*

Mean

The mean, or average, is the sum of all the elements divided by the number of elements in the distribution. It is symbolized by μ in a population and by $\overline{X}$ ("x-bar") in a sample. The formulae for calculating the mean are therefore

$$\mu = \frac{\Sigma X}{N}\text{ in a population and }\overline{X} = \frac{\Sigma X}{n}\text{ in a sample,}$$

where Σ is "the sum of" so that $\Sigma X = X_1 + X_2 + X_3 + \ldots X_n$

Unlike other measures of central tendency, the mean responds to the exact value of every score in the distribution, and unlike the median and the mode, it is very sensitive to extreme scores. As a result, it is usually an inappropriate measure for characterizing very skewed distributions. On the other hand, it has a desirable property: repeated samples drawn from the same population will tend to have very similar means, and so the mean is the measure of central tendency that best resists the influence of fluctuation between different samples. For example, if repeated blood samples were taken from a patient, the mean number of white blood cells per high-powered microscope field would fluctuate less from sample to sample than would the modal or median number of cells.

The relationship among the three measures of central tendency depends on the shape of the distribution. In a unimodal symmetrical distribution (such as the normal distribution), all three measures are identical, but in a skewed distribution, they will usually differ. Figures 1-8 and 1-9 show positively and negatively skewed distributions, respectively. In both of these, the mode is simply the most frequently occurring score (the highest point on the curve); the mean is pulled up or down by the influence of a relatively small number of very high or very low scores; and the median lies between the two, dividing the distribution into two equal areas under the curve.

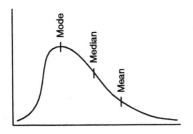

● **Figure 1-8** Measures of central tendency in a positively skewed distribution.

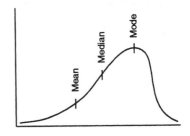

● **Figure 1-9** Measures of central tendency in a negatively skewed distribution.

Measures of Variability

Figure 1-10 shows two normal distributions, A and B; their means, modes, and medians are all identical, and, like all normal distributions, they are symmetrical and unimodal. Despite these similarities, these two distributions are obviously different; therefore, describing a normal distribution in terms of the three measures of central tendency alone is clearly inadequate.

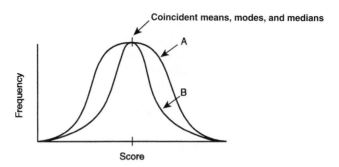

● **Figure 1-10** Normal distributions with identical measures of central tendency but different variabilities.

Although these two distributions have identical measures of central tendency, they differ in terms of their **variability**—the extent to which their scores are clustered together or scattered about. The scores forming distribution A are clearly more scattered than are those forming distribution B. Variability is a very important quality: if these two distributions represented the fasting glucose levels of diabetic patients taking two different drugs for glycemic control, for example, then drug B would be the better medication, as fewer patients on this distribution have very high or very low glucose levels—even though the *mean* effect of drug B is the same as that of drug A.

There are three important measures of variability: **range**, **variance**, and **standard deviation**.

RANGE

The range is the simplest measure of variability. It is the difference between the highest and the lowest scores in the distribution. It therefore responds to these two scores only.

For example, in the distribution 6, 9, 15, 17, 24, the range is $(24 - 6) = 18$, but in the distribution 6, 9, 15, 17, 24, 500, the range is $(500 - 6) = 494$.

VARIANCE (AND DEVIATION SCORES)

Calculating variance (and standard deviation) involves the use of **deviation scores**. The deviation score of an element is found by subtracting the distribution's mean from the element. A deviation score is symbolized by the letter x (as opposed to X, which symbolizes an element); so the formula for deviation scores is as follows:

$$x = X - \overline{X}$$

For example, in a distribution with a mean of 16, an element of 23 would have a deviation score of $(23 - 16) = 7$. On the same distribution, an element of 11 would have a deviation score of $(11 - 16) = -5$.

When calculating deviation scores for all the elements in a distribution, the results can be verified by checking that the sum of the deviation scores for all the elements is zero, that is, $\Sigma x = 0$.

The **variance** of a distribution is the mean of the squares of all the deviation scores in the distribution. The variance is therefore obtained by

- finding the deviation score (x) for each element,
- squaring each of these deviation scores (thus eliminating minus signs), and then
- obtaining their mean in the usual way—by adding them all up and then dividing the total by their number

Population variance is symbolized by σ^2. Thus,

$$\sigma^2 = \frac{\Sigma(X - \mu)^2}{N} \text{ or } \frac{\Sigma x^2}{N}$$

Sample variance is symbolized by S^2. It is found using a similar formula, but the denominator used is $n - 1$ rather than n:

$$S^2 = \frac{\Sigma(X - \overline{X})^2}{n - 1} \text{ or } \frac{\Sigma x^2}{n - 1}$$

The reason for this is somewhat complex and is not within the scope of this book or of USMLE; in practice, using $n - 1$ as the denominator gives a less-biased estimate of the variance of the population than using a denominator of n, and using $n - 1$ in this way is the generally accepted formula.

Variance is sometimes known as **mean square**. Variance is expressed in squared units of measurement, limiting its usefulness as a descriptive term—its intuitive meaning is poor.

STANDARD DEVIATION

The standard deviation remedies this problem: it is the *square root* of the variance, so it is expressed in the same units of measurement as the original data. The symbols for standard deviation are therefore the same as the symbols for variance, but without being raised to the power of two, so the standard deviation of a population is σ and the standard deviation of a sample is S. Standard deviation is sometimes written as *SD*.

The standard deviation is particularly useful in normal distributions because the proportion of elements in the normal distribution (i.e., the proportion of the area under the curve) is a constant for a given number of standard deviations above or below the mean of the distribution, *as shown in Figure 1-11.*

In Figure 1-11:
* Approximately 68% of the distribution falls within ±1 standard deviations of the mean.
* Approximately 95% of the distribution falls within ±2 standard deviations of the mean.
* Approximately 99.7% of the distribution falls within ±3 standard deviations of the mean.

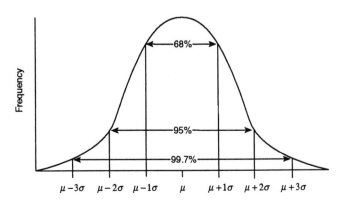

● **Figure 1-11** Standard deviation and the proportion of elements in the normal distribution.

Because these proportions hold true for every normal distribution, they should be memorized.

Therefore, if a population's resting heart rate is normally distributed with a mean (μ) of 70 and a standard deviation (S) of 10, the proportion of the population that has a resting heart rate between certain limits can be stated.

As Figure 1-12 shows, because 68% of the distribution lies within approximately ±1 standard deviations of the mean, 68% of the population will have a resting heart rate between 60 and 80 beats/min.

Similarly, 95% of the population will have a heart rate between approximately $70 \pm (2 \times 10) = 50$ and 90 beats/min (i.e., within 2 standard deviations of the mean).

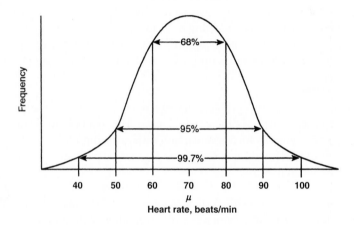

● **Figure 1-12** The normal distribution of heart rate in a hypothetical population.

Z Scores

The location of any element in a normal distribution can be expressed in terms of how many standard deviations it lies above or below the mean of the distribution. This is the *z score* of the element. If the element lies above the mean, it will have a positive *z* score; if it lies below the mean, it will have a negative *z* score.

For example, a heart rate of 85 beats/min in the distribution shown in Figure 1-12 lies 1.5 standard deviations above the mean, so it has a *z* score of +1.5. A heart rate of 65 lies 0.5 standard deviations below the mean, so its *z* score is −0.5. The formula for calculating *z* scores is therefore

$$z = \frac{X - \mu}{\sigma}$$

TABLES OF *Z* SCORES

Tables of *z* scores state what proportion of any normal distribution lies above or below *any* given *z* scores, not just *z* scores of ±1, 2, or 3.

Table 1-3 is an abbreviated table of *z* scores; it shows, for example, that 0.3085 (or about 31%) of any normal distribution lies above a *z* score of +0.5. Because normal distributions are symmetrical, this also means that approximately 31% of the distribution lies *below* a *z* score of −0.5 (which

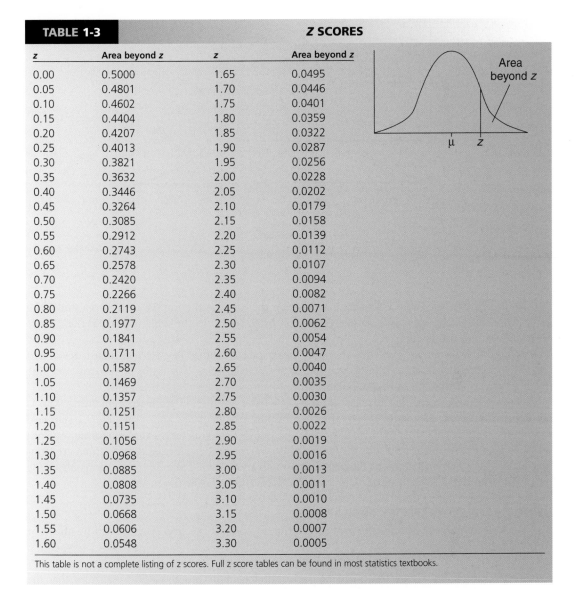

z	Area beyond z	z	Area beyond z
0.00	0.5000	1.65	0.0495
0.05	0.4801	1.70	0.0446
0.10	0.4602	1.75	0.0401
0.15	0.4404	1.80	0.0359
0.20	0.4207	1.85	0.0322
0.25	0.4013	1.90	0.0287
0.30	0.3821	1.95	0.0256
0.35	0.3632	2.00	0.0228
0.40	0.3446	2.05	0.0202
0.45	0.3264	2.10	0.0179
0.50	0.3085	2.15	0.0158
0.55	0.2912	2.20	0.0139
0.60	0.2743	2.25	0.0112
0.65	0.2578	2.30	0.0107
0.70	0.2420	2.35	0.0094
0.75	0.2266	2.40	0.0082
0.80	0.2119	2.45	0.0071
0.85	0.1977	2.50	0.0062
0.90	0.1841	2.55	0.0054
0.95	0.1711	2.60	0.0047
1.00	0.1587	2.65	0.0040
1.05	0.1469	2.70	0.0035
1.10	0.1357	2.75	0.0030
1.15	0.1251	2.80	0.0026
1.20	0.1151	2.85	0.0022
1.25	0.1056	2.90	0.0019
1.30	0.0968	2.95	0.0016
1.35	0.0885	3.00	0.0013
1.40	0.0808	3.05	0.0011
1.45	0.0735	3.10	0.0010
1.50	0.0668	3.15	0.0008
1.55	0.0606	3.20	0.0007
1.60	0.0548	3.30	0.0005

TABLE 1-3   **Z SCORES**

This table is not a complete listing of z scores. Full z score tables can be found in most statistics textbooks.

corresponds to a heart rate of 65 beats/min in Fig. 1-12)—so approximately 31% of this population has a heart rate below 65 beats/min. By subtracting this proportion from 1, it is apparent that 0.6915, or about 69%, of the population has a heart rate of *above* 65 beats/min.

Z scores are standardized or normalized, so they allow scores on different normal distributions to be compared. For example, a person's height could be compared with his or her weight by means of his or her respective z scores (provided that both these variables are elements in normal distributions).

Instead of using z scores to find the proportion of a distribution corresponding to a particular score, we can also do the converse: use z scores to find the score that divides the distribution into specified proportions.

For example, if we want to know what heart rate divides the fastest-beating 5% of the population (i.e., the group at or above the 95th percentile) from the remaining 95%, we can use the z score table.

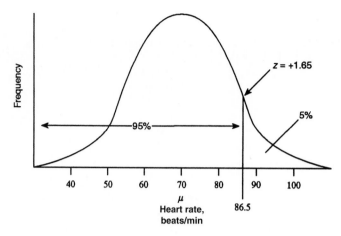

● **Figure 1-13** Heart rate of the fastest-beating 5% of the population.

To do this, we use Table 1-3 to find the z score that divides the top 5% of the area under the curve from the remaining area. The nearest figure to 5% (0.05) in the table is 0.0495; the z score corresponding to this is 1.65.

As Figure 1-13 shows, the corresponding heart rate therefore lies 1.65 standard deviations above the mean; that is, it is equal to $\mu + 1.65\sigma = 70 + (1.65 \times 10) = 86.5$. We can conclude that the fastest-beating 5% of this population has a heart rate above 86.5 beats/min.

Note that the z score that divides the top 5% of the population from the remaining 95% is *not* approximately 2. Although 95% of the distribution falls between approximately ±2 standard deviations of the mean, this is the *middle* 95% (see Fig. 1-12). This leaves the remaining 5% split into two equal parts at the two tails of the distribution (remember, normal distributions are symmetrical). Therefore, only 2.5% of the distribution falls more than 2 standard deviations *above* the mean, and another 2.5% falls more than 2 standard deviations *below* the mean.

USING *Z* SCORES TO SPECIFY PROBABILITY

Z scores also allow us to specify the probability that a randomly picked element will lie above or below a particular score.

> For example, if we know that 5% of the population has a heart rate above 86.5 beats/min, then the probability of one randomly selected person from this population having a heart rate above 86.5 beats/min will be 5%, or .05.

We can find the probability that a random person will have a heart rate less than 50 beats/min in the same way. Because 50 lies 2 standard deviations (i.e., 2×10) below the mean (70), it corresponds to a z score of -2, and we know that approximately 95% of the distribution lies within the limits $z = \pm 2$. Therefore, 5% of the distribution lies outside these limits, equally in each of the two tails of the distribution. 2.5% of the distribution therefore lies below 50, so the probability that a randomly selected person has a heart rate less than 50 beats/min is 2.5%, or .025.

Chapter 2

Inferential Statistics

At the end of the previous chapter, we saw how z scores can be used to find the probability that a random element will have a score above or below a certain value. To do this, the population had to be normally distributed, and both the population mean (μ) and the population standard deviation (σ) had to be known.

Most research, however, involves the opposite kind of problem: instead of using information about a *population* to draw conclusions or make predictions about a *sample,* the researcher usually wants to use the information provided by a *sample* to draw conclusions about a *population.* For example, a researcher might want to forecast the results of an election on the basis of an opinion poll, or predict the effectiveness of a new drug for all patients with a particular disease after it has been tested on only a small sample of patients.

Statistics and Parameters

In such problems, the population mean and standard deviation, μ and σ (which are called the population **parameters**), are unknown; all that is known is the sample mean ($\overline{X}$) and standard deviation (S)—these are called the sample **statistics**. The task of using a sample to draw conclusions about a population involves going beyond the actual information that is available; in other words, it involves **inference**. Inferential statistics therefore involve using a statistic to estimate a parameter.

However, it is unlikely that a sample will perfectly represent the population it is drawn from: a statistic (such as the sample mean) will not exactly reflect its corresponding parameter (the population mean). For example, in a study of intelligence, if a sample of 1,000 people is drawn from a population with a mean IQ of 100, it would not be expected that the mean IQ of the sample would be *exactly* 100. There will be **sampling error**—which is not an error, but just natural, expected random variation—that will cause the sample statistic to differ from the population parameter. Similarly, if a coin is tossed 1,000 times, even if it is perfectly fair, we would not expect to get *exactly* 500 heads and 500 tails.

THE RANDOM SAMPLING DISTRIBUTION OF MEANS

Imagine you have a hat containing 100 cards, numbered from 0 to 99. At random, you take out five cards, record the number written on each one, and find the mean of these five numbers. Then you put the cards back in the hat and draw another random sample, repeating the same process for about 10 minutes.

Do you expect that the means of each of these samples will be exactly the same? Of course not. Because of sampling error, they vary somewhat. If you plot all the means on a frequency distribution, the sample means form a distribution, called the **random sampling distribution of means**. If you actually try this, you will note that this distribution looks pretty much like a normal distribution. If you continued drawing samples and plotting their means *ad infinitum,* you would find that the distribution actually becomes a normal distribution! This holds true even if the underlying population was not at all normally distributed: in our population of cards in the hat, there is just one card with each number, so the shape of the distribution is actually *rectangular,* as shown in Figure 2-1, yet its random sampling distribution of means still tends to be normal.

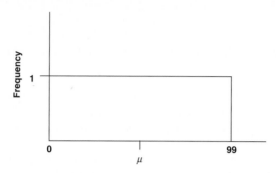

● **Figure 2-1** Distribution of population of 100 cards, each marked with a unique number between 0 and 99.

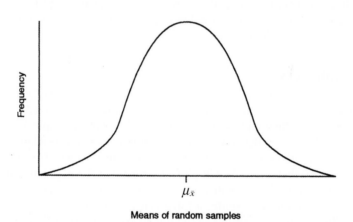

Means of random samples

● **Figure 2-2** The random sampling distribution of means: the ultimate result of drawing a large number of random samples from a population and plotting each of their individual means on a frequency distribution.

These principles are stated by the **central limit theorem**, which states that *the random sampling distribution of means will always tend to be normal, irrespective of the shape of the population distribution from which the samples were drawn.* Figure 2-2 is a random sampling distribution of means; even if the underlying population formed a rectangular, skewed, or any other nonnormal distribution, the means of all the random samples drawn from it will always tend to form a normal distribution. The theorem further states that *the random sampling distribution of means will become closer to normal as the size of the samples increases.*

According to the theorem, the mean of the random sampling distribution of means (symbolized by $\mu_{\bar{x}}$, showing that it is the mean of the population of all the sample means) is equal to the mean of the original population; in other words, $\mu_{\bar{x}}$ is equal to μ. (If Figure 2-2 were superimposed on Figure 2-1, the means would be the same).

Like all distributions, the random sampling distribution of means shown in Figure 2-2 not only has a mean, but also has a standard deviation. As always, standard deviation is a measure of variability—a measure of the degree to which the elements of the distribution are clustered together or scattered widely apart. This particular standard deviation, the standard deviation of the random sampling distribution of means, is symbolized by $\sigma_{\bar{x}}$, signifying that it is the standard deviation of the population of all the sample means. It has its own name: **standard error**, or **standard error of the mean**, sometimes abbreviated as *SE* or *SEM*. It is a measure of the extent to which the sample means deviate from the true population mean.

Figure 2-2 shows the obvious: when repeated random samples are drawn from a population, most of the means of those samples are going to cluster around the original population mean. In the "numbers in the hat" example, we would expect to find many sample means clustering around 50 (say, between 40 and 60). Rather fewer sample means would fall between 30 and 40 or between 60 and 70. Far fewer would lie out toward the extreme "tails" of the distribution (between 0 and 20 or between 80 and 99).

If the samples each consisted of just two cards what would happen to the shape of Figure 2-2? Clearly, with an *n* of just 2, there would be quite a high chance of any particular sample mean falling out toward the tails of the distribution, giving a broader, fatter shape to the curve, and hence a higher standard error. On the other hand, if the samples consisted of 25 cards each ($n = 25$), it would be very unlikely for many of their means to lie far from the center of the curve. Therefore, there would be a much thinner, narrower curve and a lower standard error.

Thus, the shape of the random sampling distribution of means, as reflected by its standard error, is affected by the size of the samples. In fact, the standard error is equal to the population standard deviation (σ) divided by the square root of the size of the samples (*n*). Therefore, the formula for the standard error is

$$\sigma_{\bar{x}} = \frac{\sigma}{\sqrt{n}}$$

STANDARD ERROR

As the formula shows, the standard error is dependent on the size of the samples: *standard error is inversely related to the square root of the sample size,* so that the larger *n* becomes, the more closely will the sample means represent the true population mean. *This is the mathematical reason why the results of large studies or surveys are more trusted than the results of small ones*—a fact that is intuitively obvious!

PREDICTING THE PROBABILITY OF DRAWING SAMPLES WITH A GIVEN MEAN

Because the random sampling distribution of means is by definition normal, the known facts about normal distributions and *z* scores can be used to find the probability that a *sample* will have a *mean* of above or below a given value, provided, of course, that the sample is a random one. This is a step beyond what was possible in Chapter 1, where we could only predict the probability that *one element* would have a score above or below a given value.

In addition, because the random sampling distribution of means is normal even when the underlying population is not normally distributed, *z* scores can be used to make predictions, regardless of the underlying population distribution—provided, once again, that the sample is random.

USING THE STANDARD ERROR

The method used to make a prediction about a sample mean is similar to the method used in Chapter 1 to make a prediction about a single element—it involves finding the *z* score corresponding to the value of interest. However, instead of calculating the *z* score in terms of the number of *standard deviations* by which a given *single element* lies above or below the population mean, the *z* score is now calculated in terms of the number of *standard errors* by which a *sample mean* lies above or below the population mean. Therefore, the previous formula

$$z = \frac{X - \mu}{\sigma} \quad \text{now becomes} \quad z = \frac{\bar{X} - \mu}{\sigma_{\bar{x}}}$$

For example, in a population with a mean resting heart rate of 70 beats/min and a standard deviation of 10, the probability that a random sample of 25 people will have a mean heart rate above 75 beats/min can be determined. The steps are as follows:

1. Calculate the standard error: $\sigma_{\bar{x}} = \dfrac{\sigma}{\sqrt{n}} = \dfrac{10}{\sqrt{25}} = 2$

2. Calculate the z score of the sample mean: $z = \dfrac{\overline{X} - \mu}{\sigma_{\bar{x}}} = \dfrac{75 - 70}{2} = 2.5$

3. Find the proportion of the normal distribution that lies beyond this z score (2.5). Table 1-3 shows that this proportion is .0062. Therefore, the probability that a random sample of 25 people from this population will have a mean resting heart rate above 75 beats/min is .0062.

Conversely, it is possible to find what random sample mean ($n = 25$) is so high that it would occur in only 5% or less of all samples (in other words, what mean is so high that the probability of obtaining it is .05 or less?):

Table 1-3 shows that the z score that divides the bottom 95% of the distribution from the top 5% is 1.65. The corresponding heart rate is $\mu + 1.65\sigma_{\bar{x}}$ (the population mean plus 1.65 standard errors). As the population mean is 70 and the standard error is 2, the heart rate will be 70 + (1.65 × 2), or 73.3. Figure 2-3 shows the relevant portions of the random sampling distribution of means; the appropriate z score is +1.65, not +2, because it refers to the *top .05* of the distribution, not the top .025 and the bottom .025 together.

It is also possible to find the limits between which 95% of all possible random sample means would be expected to fall. As with any normal distribution, 95% of the random sampling distribution of means lies within approximately ±2 standard errors of the population mean (in other words, within $z = \pm 2$); therefore, 95% of all possible sample means must lie within approximately ±2 standard errors of the population mean. [As Table 1-3 shows, the *exact* z scores that correspond to the middle 95% of any normal distribution are in fact ±1.96, not ±2; the exact limits are therefore 70 ± (1.96 × 2) = 66.08 and 73.92.] Applying this to the distribution of resting heart rate, it is apparent that 95% of all possible random sample means will fall between the limits of $\mu \pm 2\sigma_{\bar{x}}$, that is, approximately 70 ± (2 × 2), or 66 and 74.

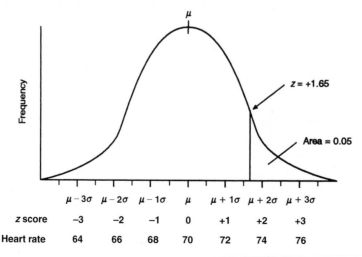

● **Figure 2-3** Mean heart rates of random samples ($n = 25$) drawn from a population with a mean heart rate of 70 and a standard deviation of 10.

Estimating the Mean of a Population

So far, we have seen how z scores are used to find the probability that a random sample will have a mean of above or below a given value. It has been shown that 95% of all possible members of the population will lie within approximately ±2 (or, more exactly, ±1.96) standard errors of the population mean and 95% of all such means will be within ±2 standard errors of the mean.

CONFIDENCE LIMITS

The sample mean ($\overline{X}$) lies within ±1.96 standard errors of the population mean (μ) 95% (.95) of the time; conversely, μ lies within ±1.96 standard errors of $\overline{X}$ 95% of the time. These limits of ±1.96 standard errors are called the **confidence limits** (in this case, the 95% confidence limits). Finding the confidence limits involves inferential statistics, because a sample statistic ($\overline{X}$) is being used to estimate a population parameter (μ).

For example, if a researcher wishes to find the true mean resting heart rate of a large population, it would be impractical to take the pulse of every person in the population. Instead, he or she would draw a random sample from the population and take the pulse of the persons in the sample. As long as the sample is truly random, the researcher can be 95% confident that the true population mean lies within ±1.96 standard errors of the sample mean.

Therefore, if the mean heart rate of the sample ($\overline{X}$) is 74 and $\sigma_{\overline{x}}$ = 2, the researcher can be 95% certain that μ lies within 1.96 standard errors of 74, i.e., between 74 ± (1.96 × 2), or 70.08 and 77.92. The best *single* estimate of the population mean is still the sample mean, 74—after all, it is the only piece of actual data on which an estimate can be based.

In general, confidence limits are equal to the sample mean plus or minus the z score obtained from the table (for the appropriate level of confidence) multiplied by the standard error:

$$\text{Confidence limits} = \overline{X} \pm z\,\sigma_{\overline{x}}$$

Therefore, 95% confidence limits (which are the ones conventionally used in medical research) are approximately equal to the sample mean plus or minus two standard errors.

The difference between the upper and lower confidence limits is called the **confidence interval**—*sometimes abbreviated as* **CI**.

Researchers obviously want the confidence interval to be as narrow as possible. The formula for confidence limits shows that to make the confidence interval narrower (for a given level of confidence, such as 95%), the standard error ($\sigma_{\overline{x}}$) must be made smaller. Standard error is found by the formula $\sigma_{\overline{x}} = \sigma \div \sqrt{n}$. Because σ is a population parameter that the researcher cannot change, the only way to reduce standard error is to increase the sample size n. Once again, there is a mathematical reason why large studies are trusted more than small ones! Note that the formula for standard error means that standard error will decrease only in proportion to the *square root* of the sample size; therefore, the width of the confidence interval will decrease in proportion to the square root of the sample size. In other words, to *halve* the confidence interval, the sample size must be increased *fourfold*.

PRECISION AND ACCURACY

Precision is *the degree to which a figure (such as an estimate of a population mean) is immune from random variation.* The width of the confidence interval reflects precision—the wider the confidence interval, the less precise the estimate.

Because the width of the confidence interval decreases in proportion to the square root of sample size, precision is proportional to the square root of sample size. So to double the precision of an estimate, sample size must be multiplied by 4; to triple precision, sample size must be multiplied by 9; and to quadruple precision, sample size must be multiplied by 16.

Increasing the precision of research therefore requires disproportionate increases in sample size; thus, very precise research is expensive and time-consuming.

Precision must be distinguished from **accuracy**, which is the degree to which an estimate is immune from systematic error or bias.

A good way to remember the difference between precision and accuracy is to think of a person playing darts, aiming at the bull's-eye in the center of the dartboard. Figure 2-4A shows how the dartboard looks after a player has thrown five darts. Is there much systematic error (bias)? No. The darts do not tend to err consistently in any one direction. However, although there is no bias, there is much random variation, as the darts are not clustered together. Hence, the player's aim is unbiased (or accurate) but imprecise. It may seem strange to call such a poor player accurate, but the darts are at least centered on the bull's-eye, on average. The player needs to reduce the random variation in his or her aim, rather than aim at a different point.

Figure 2-4B shows a different situation. Is there much systematic error or bias? Certainly. The player consistently throws toward the top left of the dartboard, and so the aim is biased (or inaccurate). Is there much random variation? No. The darts are tightly clustered together, hence relatively immune from random variation. The player's aim is therefore precise.

Figure 2-4C shows darts that are not only widely scattered, but also systematically err in one direction. Thus, this player's aim is not immune from either bias or random variation, making it biased (inaccurate) and imprecise.

Figure 2-4D shows the ideal, both in darts and in inferential statistics. There is no systematic error or significant random variation, so this aim is both accurate (unbiased) and precise.

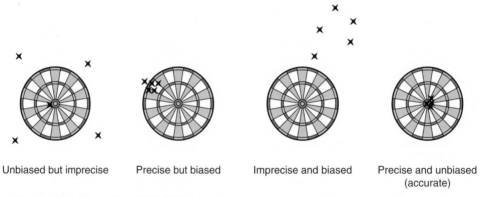

Unbiased but imprecise   Precise but biased   Imprecise and biased   Precise and unbiased (accurate)

● **Figure 2-4** Dartboard illustration of precision and accuracy

Figure 2-5 shows the same principles in terms of four hypothetical random sampling distributions of means. Each curve shows the result of taking a very large number of samples from the same population and then plotting their means on a frequency distribution. Precision is shown by the narrowness of each curve: as in all frequency distributions, the spread of the distribution around its mean reflects its variability. A very spread-out curve has a high variability and a high standard error and therefore provides an imprecise estimate of the true population mean. Accuracy is shown by the distance between the mean of the random sampling distribution of means ($\mu_{\bar{x}}$) and the true population mean (μ). This is analogous to a darts player with an inaccurate aim and a considerable distance between the average position of his or her darts and the bull's-eye.

Distribution A in Figure 2-5 is a very spread-out random sampling distribution of means; thus, it provides an imprecise estimate of the true population mean. However, its mean does coincide with the true population mean, and so it provides an accurate estimate of the true population mean. In other words, the estimate that it provides is not biased, but it is subject to considerable random variation. This is the type of result that would occur if the samples were truly random but small.

Distribution B is a narrow distribution, which therefore provides a precise estimate of the true population mean. Due to the low standard error, the width of the confidence interval would be narrow. However, its mean lies a long way from the true population mean, so it will provide a

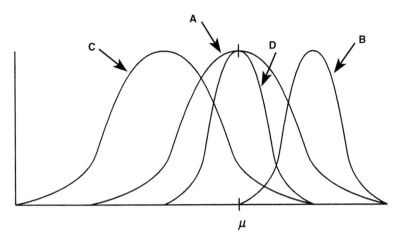

● Figure 2-5 Four hypothetical random sampling distributions of means.

biased estimate of the true population mean. This is the kind of result that is produced by large but biased (i.e., not truly random) samples.

Distribution C has the worst of both worlds: it is very spread out (having a high standard error) and would therefore provide an imprecise estimate of the true population mean. Its mean lies a long way from the true population mean, so its estimate is also biased. This would occur if the samples were small and biased.

Distribution D is narrow, and therefore precise, and its mean lies at the same point as the true population mean, so it is also accurate. This ideal is the kind of distribution that would be obtained from large and truly random samples; therefore, to achieve maximum precision and accuracy in inferential statistics, samples should be large and truly random.

ESTIMATING THE STANDARD ERROR

So far, it has been shown how to determine the probability that a random sample will have a mean that is above or below a certain value, and we have seen how the mean of a sample can be used to estimate the mean of the population from which it was drawn, with a known degree of precision and confidence. All this has been done by using z scores, which express the number of standard errors by which a sample mean lies above or below the true population mean.

However, because standard error is found from the formula $\sigma_{\bar{x}} = \sigma \div \sqrt{n}$, we cannot calculate standard error unless we know σ, the population standard deviation. In practice, σ will not be known: researchers hardly ever know the standard deviation of the population (and if they did, they would probably not need to use inferential statistics anyway).

As a result, standard error cannot be calculated, and so z scores cannot be used. However, the standard error can be *estimated* using data that *are* available from the sample alone. The resulting statistic is the **estimated standard error of the mean**, usually called estimated standard error (although, confusingly, it is called standard error in many research articles); it is symbolized by $s_{\bar{x}}$, and it is found by the formula

$$s_{\bar{x}} = \frac{S}{\sqrt{n}}$$

where S is the sample standard deviation, as defined in Chapter 1.

t SCORES

The estimated standard error is used to find a statistic, t, that can be used in place of z. The t score, rather than the z score, *must* be used when making inferences about means that are based on *estimates* of population parameters (such as estimated standard error) rather than on the population parameters themselves. The t score is sometimes known as **Student's** t; it is calculated in much the same way as z. But while z was expressed in terms of the number of *standard errors* by which a sample mean lies above or below the population mean, t is expressed in terms of the number of *estimated* standard errors by which the sample mean lies above or below the population mean. The formula for t is therefore

$$t = \frac{\overline{X} - \mu}{s_{\overline{x}}}$$

Compare this formula with the formula we used for z:

$$z = \frac{\overline{X} - \mu}{\sigma_{\overline{x}}}$$

Just as z score tables give the proportions of the normal distribution that lie above and below any given z score, t score tables provide the same information for any given t score. However, there is one difference: while the value of z for any given proportion of the distribution is constant (e.g., z scores of ±1.96 *always* delineate the middle 95% of the distribution), the value of t for any given proportion is not constant—it varies according to sample size. When the sample size is large ($n > 100$), the values of t and z are similar, but as samples get smaller, t and z scores become increasingly different.

DEGREES OF FREEDOM AND t TABLES

Table 2-1 is an abbreviated t score table that shows the values of t corresponding to different areas under the normal distribution for various sample sizes. Sample size (n) is not stated directly in t score tables; instead, the tables express sample size in terms of **degrees of freedom** (*df*). The mathematical concept behind degrees of freedom is complex and not needed for the purposes of USMLE or understanding statistics in medicine: for present purposes, df can be defined as simply equal to $n - 1$. Therefore, to determine the values of t that delineate the central 95% of the sampling distribution of means based on a sample size of 15, we would look in the table for the appropriate value of t for $df = 14$ (14 being equal to $n - 1$); this is sometimes written as t_{14}. Table 2-1 shows that this value is 2.145.

As n becomes larger (100 or more), the values of t are very close to the corresponding values of z. As the middle column shows, for a df of 100, 95% of the distribution falls within $t = \pm1.984$, while for a df of ∞, this figure is 1.96, which is the same as the figure for z (see Table 1-3). In general, the value of t that divides the central 95% of the distribution from the remaining 5% is in the region of 2, just as it is for z. (One- and two-tailed tests will be discussed in Chapter 3.)

As an example of the use of t scores, we can repeat the earlier task of estimating (with 95% confidence) the true mean resting heart rate of a large population, basing the estimate on a random sample of people drawn from this population. This time we will not make the unrealistic assumption that the standard error is known.

As before, a random sample of 15 people is drawn; their mean heart rate ($\overline{X}$) is 74 beats/min. If we find that the standard deviation of this sample is 8.2, the estimated standard error, $s_{\overline{x}}$, can be calculated as follows:

$$s_{\overline{x}} = \frac{s}{\sqrt{n}}$$
$$= \frac{8.2}{\sqrt{15}}$$
$$= \frac{8.2}{3.87}$$
$$= 2.1$$

TABLE 2-1	ABBREVIATED TABLE OF *t* SCORES			
Area in 2 tails	.100	.050	.010	Tail 1 Tail 2
Area in 1 tail	.050	.025	.005	Tail 1
df				
1	6.314	12.706	63.657	
2	2.920	4.303	9.925	
3	2.353	3.182	5.841	
4	2.132	2.776	4.604	
5	2.015	2.571	4.032	
6	1.943	2.447	3.707	
7	1.895	2.365	3.499	
8	1.860	2.306	3.355	
9	1.833	2.262	3.250	
10	1.812	2.228	3.169	
11	1.796	2.201	3.106	
12	1.782	2.179	3.055	
13	1.771	2.160	3.012	
14	1.761	2.145	2.977	
15	1.753	2.131	2.947	
25	1.708	2.060	2.787	
50	1.676	2.009	2.678	
100	1.660	1.984	2.626	
∞	1.645	1.960	2.576	

This table is not a complete listing of *t*-statistics values. Full tables may be found in most statistics textbooks.

For a sample consisting of 15 people, the *t* tables will give the appropriate value of *t* (corresponding to the middle 95% of the distribution) for $df = 14$ (i.e., $n - 1$).

Table 2-1 shows that this value is 2.145. This value is not very different from the "ballpark" 95% figure for z, which is 2. The 95% confidence intervals are therefore equal to the sample mean plus or minus *t* times the estimated standard error (i.e., $\bar{X} \pm t \times s_{\bar{x}}$), which in this example is

$$74 \pm (2.145 \times 2.1) = 69.5 \text{ and } 78.5.$$

The sample mean therefore allows us to estimate that the true mean resting heart rate of this population is 74 beats/min, and we can be 95% confident that it lies between 69.5 and 78.5.

Note that in general, one can be 95% confident that the true mean of a population lies within approximately plus or minus two estimated standard errors of the mean of a random sample drawn from that population.

Chapter 3

Hypothesis Testing

Chapter 2 showed how a statistic (such as the mean of a sample) can be used to estimate a parameter (such as the mean of a population) with a known degree of confidence. This is an important use of inferential statistics, but a more important use is *hypothesis testing*.

Hypothesis testing may seem complex at first, but the steps involved are actually very simple. To test a hypothesis about a mean, the steps are as follows:

1. State the null and alternative hypotheses, H_0 and H_A.
2. Select the decision criterion α (or "level of significance").
3. Establish the critical values.
4. Draw a random sample from the population, and calculate the mean of that sample.
5. Calculate the standard deviation (S) and estimated standard error of the sample ($s_{\bar{x}}$).
6. Calculate the value of the test statistic t that corresponds to the mean of the sample (t_{calc}).
7. Compare the calculated value of t with the critical values of t, and then accept or reject the null hypothesis.

Step 1: State the Null and Alternative Hypotheses

Consider the following example. The Dean of a medical school states that the school's students are a highly intelligent group with an average IQ of 135. This claim is a hypothesis that can be tested; it is called the **null hypothesis**, or H_0. It has this name because in most research it is the hypothesis of no difference between samples or populations being compared (e.g., that a new drug produces no change compared with a placebo). If this hypothesis is rejected as false, then there is an **alternative (or experimental) hypothesis**, H_A, that logically must be accepted. In the case of the Dean's claim, the following hypotheses can be stated:

$$\text{Null hypothesis, } H_0\text{: } \mu = 135$$

$$\text{Alternative hypothesis, } H_A\text{: } \mu \neq 135$$

One way of testing the null hypothesis would be to measure the IQ of every student in the school—in other words, to test the entire population—but this would be expensive and time-consuming. It would be more practical to draw a random sample of students, find their mean IQ, and then make an inference from this sample.

Step 2: Select the Decision Criterion α

If the null hypothesis were correct, would the mean IQ of the sample of students be expected to be exactly 135?

No, of course not. As shown in Chapter 2, sampling error will always cause the mean of the sample to deviate from the mean of the population. For example, if the mean IQ of the sample were 134, we might reasonably conclude that the null hypothesis was not contradicted because sampling error could easily permit a sample with this mean to have been drawn from a population with a mean of 135. To reach a conclusion about the null hypothesis, we must therefore decide *at what point is the difference between the sample mean and 135 not due to chance* but due to the fact that the population mean is *not* really 135, as the null hypothesis claims?

This point must be set before the sample is drawn and the data are collected. Instead of setting it in terms of the actual IQ score, it is set in terms of probability. The probability level at which it is decided that the null hypothesis is incorrect constitutes a **criterion**, or significance level, known as α (alpha).

As the random sampling distribution of means (Fig. 2-2) showed, it is unlikely that a random sample mean will be *very* different from the true population mean. If it is very different, lying far toward one of the tails of the curve, it arouses suspicion that the sample was *not* drawn from the population specified in the null hypothesis, but from a different population. [If a coin were tossed repeatedly and 5, 10, or 20 heads occur in a row, we would start to question the unstated assumption, or null hypothesis, that it was a fair coin (i.e., H_0: heads = tails).]

In other words, the greater the difference between the sample mean and the population mean specified by the null hypothesis, the less probable it is that the sample really does come from the specified population. When this probability is very low, we can conclude that the null hypothesis is incorrect.

How low does this probability need to be for the null hypothesis to be rejected as incorrect? By convention, the null hypothesis will be rejected if the probability that the sample mean could have come from the hypothesized population is less than or equal to .05; thus, the conventional level of α is .05. Conversely, if the probability of obtaining the sample mean is greater than .05, the null hypothesis will be accepted as correct. Although α may be set lower than the conventional .05 (for reasons which will be shown later), it may not normally be set any higher than this.

Step 3: Establish the Critical Values

In Chapter 2 we saw that if a very large number of random samples are taken from any population, their means form a normal distribution—the random sampling distribution of means—that has a mean ($\mu_{\bar{x}}$) equal to the population mean (μ). We also saw that we can specify the values of random sample means that are so high, or so low, that these means would occur in only 5% (or fewer) of all possible random samples. This ability can now be put to use because the problem of testing the null hypothesis about the students' mean IQ involves stating which random sample means are so high or so low that they would occur in only 5% (or fewer) of all random samples that could be drawn from a population with a mean of 135.

If the sample mean falls inside the range within which we would expect 95% of random sample means to fall, the null hypothesis is accepted. This range is therefore called the **area of acceptance**. If the sample mean falls outside this range, in the **area of rejection**, the null hypothesis is rejected, and the alternative hypothesis is accepted.

The limits of this range are called the **critical values**, and they are established by referring to a table of *t* scores.

In the current example, the following values can be calculated:

- The sample size is 10, so there are $(n - 1) = 9$ degrees of freedom
- The table of *t* scores (Table 2-1) shows that when $df = 9$, the value of *t* that divides the 95% (0.95) area of acceptance from the two 2.5% (0.025) areas of rejection is ±2.262. These are the critical values, which are written $t_{\text{crit}} = \pm 2.262$

Figure 3-1 shows the random sampling distribution of means for our hypothesized population with a mean (μ) of 135. It also shows the areas of rejection and acceptance defined by the critical values of *t* that were just established. As shown, the hypothesized population mean is sometimes written μ_{hyp}.

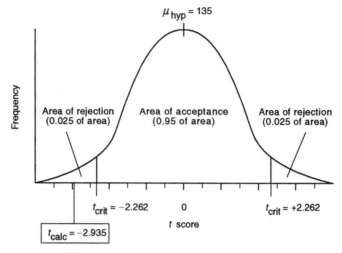

● **Figure 3-1** Random sampling distribution of means for a hypothesized population with a mean of 135.

We have now established the following:

- The null and alternative hypotheses
- The criterion that will determine when the null hypothesis will be accepted or rejected
- The critical values of *t* associated with this criterion

A random sample of students can now be drawn from the population; the *t* score (t_{calc}) associated with their mean IQ can then be calculated and compared with the critical values of *t*. This is a *t*-test—a very common test in medical literature.

Step 4: Draw a Random Sample from the Population and Calculate the Mean of That Sample

A random sample of 10 students is drawn; their IQs are as follows:

$$115\ldots140\ldots133\ldots125\ldots120\ldots126\ldots136\ldots124\ldots132\ldots129$$

The mean ($\overline{X}$) of this sample is 128.

Step 5: Calculate the Standard Deviation (S) and Estimated Standard Error of the Sample ($s_{\overline{x}}$)

To calculate the *t* score corresponding to the sample mean, the estimated standard error must first be found. This is done as described in Chapter 2. The standard deviation (*S*) of this sample is calculated and found to be 7.542. The estimated standard error ($s_{\overline{x}}$) is then calculated as follows:

$$s_{\overline{x}} = \frac{S}{\sqrt{n}}$$
$$= \frac{7.542}{\sqrt{10}}$$
$$= 2.385$$

Step 6: Calculate the Value of t That Corresponds to the Mean of the Sample (t_{calc})

Now that the estimated standard error has been determined, the t score corresponding to the sample mean can be found. It is the number of estimated standard errors by which the sample mean lies above or below the hypothesized population mean:

$$t = \frac{\overline{X} - \mu_{hyp}}{s_{\overline{x}}}$$

$$= \frac{128 - 135}{2.385}$$

$$= -2.935$$

So the sample mean (128) lies approximately 2.9 estimated standard errors below the hypothesized population mean (135).

Step 7: Compare the Calculated Value of t with the Critical Values of t, and then accept or Reject the Null Hypothesis

If the calculated value of t associated with the sample mean falls at or beyond either of the critical values, it is within one of the two areas of rejection.

Figure 3-2 shows that the t score in this example *does* fall within the lower area of rejection.

Therefore, the null hypothesis is rejected, and the alternative hypothesis is accepted.

The reasoning behind this is as follows. The sample mean differs so much from the hypothesized population mean that the probability that it would have been obtained if the null hypothesis were true is only .05 (or less). Because this probability is so low, we conclude that the population mean is not 135. We can say that the difference between the sample mean and the hypothesized population mean is **statistically significant**, and the null hypothesis is rejected at the 0.05 level. This would typically be reported as follows: "The hypothesis that the mean IQ of the population is 135 was rejected, $t = -2.935$, $df = 9$, $p \leq .05$."

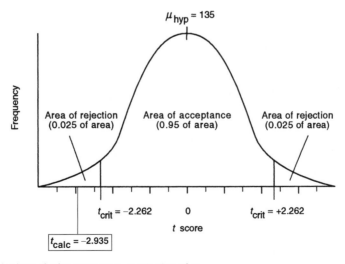

● **Figure 3-2** Critical values of **t** for acceptance or rejection of H_0.

If, on the other hand, the calculated value of *t* associated with the sample mean fell between the two critical values, in the area of acceptance, the null hypothesis would be accepted. In such a case, it would be said that the difference between the sample mean and the hypothesized population mean failed to reach statistical significance (*p* > .05).

Z-Tests

References to a "*z*-test" are sometimes made in medical literature. A *z*-test involves the same steps as a *t*-test and can be used when the sample is large enough (*n* > 100) for the sample standard deviation to provide a reliable estimate of the standard error. Although there are situations in which a *t*-test can be used but a *z*-test cannot, there are no situations in which a *z*-test can be used but a *t*-test cannot. Therefore, *t*-tests are the more important and widely used of the two.

The Meaning of Statistical Significance

When a result is reported to be "significant at p ≤ .05," it merely means that the result was unlikely to have occurred by chance—in this case, that the likelihood of the result having occurred by chance is .05 or less. This does not necessarily mean that the result is truly "significant" in the everyday meaning of the word—that it is important, noteworthy, or meaningful. Nor does it mean that it is necessarily clinically significant.

In the previous example, if the mean IQ of the sample of students were found to be 134, it is possible (if the sample were large enough) that this mean could fall in the area of rejection, and so the null hypothesis (*μ* = 135) could be rejected. However, this would scarcely be an important or noteworthy disproof of the Dean's claim about the students' intelligence.

In fact, virtually *any* null hypothesis can be rejected if the sample is sufficiently large, because there will almost always be some trivial difference between the hypothesized mean and the sample mean. Studies using extremely large samples therefore risk producing findings that are statistically significant but otherwise insignificant. For example, a study of an antihypertensive drug versus a placebo might conclude that the drug was effective—but if the difference in blood pressure was only 1 mm Hg, this would not be a significant finding in the usual meaning of the word, and would not lead physicians to prescribe the drug.

Type I and Type II Errors

A statement that a result is "significant at *p* ≤ .05" means that an investigator can be 95% sure that the result was not obtained by chance. It also means that there is a 5% probability that the result *could* have been obtained by chance. Although the null hypothesis is being rejected, it *could* still be true: there remains a 5% chance that the data *did*, in fact, come from the population specified by the null hypothesis.

Questions on types I and II errors will appear not only on Step 1, but also on Step 2, Step 3, and even specialty board certification examinations.

Accepting the alternative (or experimental) hypothesis when it is false is a **type I** or "**false positive**" error: a positive conclusion has been reached about a hypothesis that is actually false. The probability that a type I error is being made is in fact the value of *p*; because this value relates to the criterion α, a type I error is also known as an **alpha error**.

The opposite kind of error, rejecting the alternative (or experimental) hypothesis when it is true is a **type II** or "**false negative**" error: a negative conclusion has been drawn about a hypothesis that is actually true. This is also known as a **beta error**. While the probability of making a type I error is α, the probability of making a type II error is β. Table 3-1 shows the four possible kinds of decisions that can be made on the basis of statistical tests.

TABLE 3-1	THE FOUR POSSIBLE KINDS OF DECISIONS THAT CAN BE MADE ON THE BASIS OF STATISTICAL TESTS	
	ACTUAL SITUATION	
	H_0 True / H_A False	H_0 False / H_A True
TEST RESULT — H_0 Accepted / H_A Rejected	Correct	Type II error (β) False negative
H_0 Rejected / H_A Accepted	Type I error (α) False positive	Correct

The choice of an appropriate level for the criterion α therefore depends on the relative consequences of making a type I or type II error. For example, if a study is expensive and time-consuming (and is therefore unlikely to be repeated), yet has important practical implications, the researchers may wish to establish a more stringent level of α (such as .01, .005, or even .001) to be more than 95% sure that their conclusions are correct. This was done in the multimillion dollar Lipid Research Clinics Coronary Primary Prevention Trial (1979), whose planners stated that

> since the time, magnitude, and costs of this study make it unlikely that it could ever be repeated, it was essential that any observed benefit of total cholesterol lowering was a real one. Therefore, α was set to .01 rather than the usual .05.

Although the criterion to be selected need not be .05, by convention it cannot be any higher. Results that do not quite reach the .05 level of probability are sometimes reported to "approach significance" or to "show statistically significant trends," phrases that perhaps reveal the investigator's desire to find statistical significance (if no such desire was present, the result would more likely simply be reported as "nonsignificant" or "n.s.").

Many researchers do not state a predetermined criterion or report their results in terms of one; instead, they report the actual probability that the obtained result could have occurred by chance if the null hypothesis were true (e.g., "$p = .015$"). In these cases, the p value is more an "index of rarity" than a true decision criterion. The researchers are showing how unlikely it is that a type I error has been made, even though they would have still rejected the null hypothesis if the outcome were only significant at the .05 level.

Power of Statistical Tests

Although it is possible to guard against a type I error simply by using a more stringent (lower) level of α, preventing a type II error is not so easy. Because a type II error involves accepting a false null hypothesis, the ability of a statistical test to avoid a type II error depends on its ability to detect a null hypothesis that is false.

This ability is called the **power** *of the test, and it is equal to $1 - \beta$: it is the probability that a false null hypothesis will be rejected. Conventionally, a study is required to have a power of 0.8 (or a β of 0.2) to be acceptable—in other words, a study that has a less than 80% chance of detecting a false null hypothesis is generally judged to be unacceptable.*

Calculating β and determining the power of a test is complex. Nevertheless, it is clear that **a test's power, or ability to detect a false null hypothesis, will increase as**

- α **increases** (e.g., from .01 to .05). This will make the critical values of t less extreme, thus increasing the size of the areas of rejection and making rejection of the null hypothesis more likely. There will always be a trade-off between type I and type II errors: increasing α reduces the chance of a type II error, but it simultaneously increases the chance of a type I error;

- **the size of the difference between the sample mean and the hypothesized population mean increases** (this is known as the **effect size**). In the preceding example, a difference between a hypothesized population mean IQ of 135 and a sample mean IQ of 100 would be detected much more easily (and hence the null hypothesis would be rejected more easily) than a difference between a hypothesized IQ of 135 and a sample mean IQ of 128. The larger the difference, the more extreme the calculated value of t. In clinical trials, the effect size is the difference that would be clinically important, or the difference that is expected to occur between two groups in the trial—such as a difference in systolic blood pressure of 10 mm Hg between a new antihypertensive drug and a placebo. Ideally, all studies that report acceptance of the null hypothesis should also report the power of the test used so that the risk of a type II error is made clear;

- **sampling error decreases**. A lower sampling error means that the sample standard deviation (S) is reduced, which will cause the estimated standard error ($s_{\bar{x}}$) to be lower. Because t is calculated in terms of estimated standard errors, this will make the calculated value of t more extreme (whether in a positive or negative direction), increasing the likelihood that it falls in one of the areas of rejection;

- **the sample size (n) increases**. This reduces the estimated standard error ($s_{\bar{x}}$), thereby increasing the calculated value of t. Therefore, a large-scale study is more likely to detect a false null hypothesis (particularly if the effect size is small) than is a small-scale study. For example, if a coin is tossed 1,000 times, resulting in 600 heads and 400 tails, it is much easier to reject the null hypothesis (that the coin is a fair one) than if the coin is tossed 10 times and 6 heads and 4 tails are obtained.

 Increasing the sample size is the most practical and important way of increasing the power of a statistical test.

The findings of a study in which the null hypothesis is accepted may be disputed by researchers, who may argue that the study's sample was too small to detect a real difference or effect. They may replicate the study using a larger sample to improve the likelihood of getting statistically significant results that will allow them to reject the null hypothesis.

Determining the size of the sample that needs to be used is crucial; a sample that is too small may be unable to answer the research question, due to a lack of power, but one that is too large is wasteful of scarce resources. Researchers calculate the required sample size by means of a formula that incorporates the risk of a type I error (alpha, or the p value required to reject the null hypothesis, usually .05), the risk of a type II error (the power of the statistical test, usually 80% or 0.80), the variability of the data (S), and the effect size. Once again, it is clear that a larger sample is required to look for small or subtle effects than for large or obvious ones.

The concept of power can be explained by using the example of a military radar system that is being used to detect a possible impending air attack. The null hypothesis is that there are no aircraft or missiles approaching; the alternative hypothesis is that there are. Clearly, a powerful radar system is going to be more able to detect intruders than is a weak one.

What if the radar system is functioning at a very low power and the operators are not aware of this fact? They watch their screens and report that the null hypothesis is correct—there are no aircraft or missiles approaching—but the power of their system is so low that they are in great danger of making a type II, or false-negative, error. This danger is greater if the "effect size"—the difference between the presence or absence of impending attackers—is likely to be low: a single light aircraft will be detected only by a very powerful system, while a low-powered system may be adequate to detect a squadron of large bombers. So just as with a statistical test, the more subtle the phenomenon being tested for, the more powerful the test needs to be.

On the other hand, a very powerful system—like a very powerful statistical test—runs the risk of making a type I error. A phenomenon so subtle as to be trivial, such as a flock of birds, may produce a signal, which may lead the operators to reject the null hypothesis and conclude that an attack is on the way.

Directional Hypotheses

So far, we have used the example of a **nondirectional** alternative hypothesis, which merely stated that the population mean is *not* equal to 135, but did not specify whether the population mean is above or below this figure. This was appropriate because the medical school Dean claimed that the students' mean IQ was 135. His claim (which constitutes the null hypothesis) could legitimately be rejected if the sample mean IQ turned out to be significantly above *or* below 135. Therefore, as Figure 3-2 showed, there were *two* areas of rejection, one above μ_{hyp} and one below.

What if the Dean had instead claimed that the students' average IQ was at *least* 135? This claim could only be rejected if the sample mean IQ turned out to be significantly *lower* than 135. The null hypothesis is now $\mu \geq 135$, and the alternative hypothesis must now be $\mu < 135$. The alternative hypothesis is now a **directional** one, which specifies that the population mean lies in a *particular direction* with respect to the null hypothesis.

In this kind of situation, there are no longer two areas of rejection on the random sampling distribution of means. As Figure 3-3 shows, there is now only one. If α remains at .05, the area of acceptance (the area in which 95% of the means of possible samples drawn from the hypothesized population lie) now extends down from the very top end of the distribution, leaving just *one* area of rejection—the bottom 5% of the curve. The area of rejection now lies in only one tail of the distribution, rather than in both tails.

The steps involved in conducting a *t*-test of this directional null hypothesis are exactly the same as before, except that the critical value of *t* is now different. The critical value now divides the bottom 5% tail of the distribution from the upper 95%, instead of dividing the middle 95% from two tails of 2.5% each. The appropriate column of Table 2-1 shows that the new critical value of *t* (for the same *df* of 9) is -1.833, rather than the previous value of ±2.262. As Figure 3-3 shows, this new critical value is associated with only one tail of the distribution. Using this value therefore involves performing a **one-tailed** (or one-sided) statistical test, because the alternative hypothesis is directional; previously, when the alternative hypothesis was nondirectional, the test performed was a **two-tailed** (or two-sided) test.

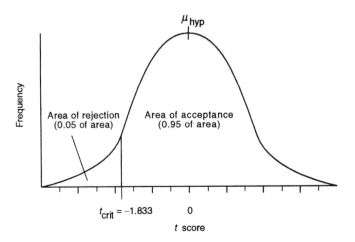

● **Figure 3-3** Areas of rejection and acceptance for a directional alternative hypothesis.

The critical value of t is less extreme for the one-tailed test (-1.833) than for the two-tailed test (± 2.262). Consequently, when a one-tailed test is used, a less extreme sample mean can exceed the critical value and falls within the area of rejection, leading to rejection of the null hypothesis. As a result of this, **one-tailed tests are more powerful than two-tailed tests**.

> For example, if the mean IQ of the sample of 10 students were 130 (instead of 128), with the same standard deviation (7.542) and the same estimated standard error (2.385) as before, the value of t corresponding to this mean would be

$$\frac{130 - 135}{2.385} = -2.096$$

> This score falls within the area of acceptance for a two-tailed test, but it falls within the area of rejection for a one-tailed test, as shown in Figure 3-3. The Dean's claim could therefore potentially be accepted *or* rejected, depending on how it is interpreted and which test is consequently performed.

As this example shows, a researcher who wishes to reject the null hypothesis may sometimes find that using a one-tailed rather than a two-tailed test allows a previously nonsignificant result to become significant. For this reason, it is important that one-tailed tests are only performed under the correct conditions. The decision to use a one-tailed test must depend on *the nature of the hypothesis being tested and should therefore be decided at the outset of the research,* rather than being decided afterward according to how the results turn out.

One-tailed tests can only be used when there is a directional alternative hypothesis. This means that they may not be used unless results in only one direction are of interest and the possibility of the results being in the opposite direction is of no interest or consequence to the researcher.

> When testing a new drug, the normal null hypothesis is that the drug has no effect, so it will be rejected if the drug turns out to have an effect too great to be due to chance, regardless of whether the effect is a positive one or a negative one. Although the researcher *expects* the drug to produce an improvement in patients' clinical state, this expectation does not permit the use of a directional alternative hypothesis. The researcher can do this only if it is of no interest or consequence if the drug actually makes patients worse, which is not usually the case in medical research. (The main exception to this is in **noninferiority trials**, which will be discussed in Chapter 5.)

Testing for Differences between Groups

We have seen how a t-test can be used to test a hypothesis about a single mean. However, biomedical research is typically more complex than this: researchers commonly want to compare *two* means, such as the effects of two different drugs or the mean survival times of patients receiving two different treatments.

A slightly more complex version of the t-test can be used to test for a significant difference between two means. The null hypothesis is that the two groups, A and B, were drawn from populations with the same mean—in other words, that the two samples were in effect drawn from the same population and that there is no difference between them. The alternative hypothesis is that the two population means are different:

$$H_0: \mu_A = \mu_B$$

$$H_A: \mu_A \neq \mu_B$$

Post Hoc Testing and Subgroup Analyses

When hypotheses exist about different groups of patients, such as according to their gender or disease severity, a **subgroup analysis** of the differences between several means will be done. For example, researchers studying a new antihypertensive medication may hypothesize that patients with high-renin levels, or obese patients, may respond better than those with low-renin levels, or nonobese patients.

When these hypotheses are prespecified, these tests are legitimate. However, it is common for researchers to test hypotheses that were *not* prespecified at the start of the research program: this is **post hoc** hypothesis testing. Post hoc subgroup analyses might appear to show that a medication significantly reduces blood pressure among men but not among women, among diabetic patients rather than nondiabetic patients, and among patients in US research centers rather than those in European ones, or perhaps that it reduces headaches, edema, or urinary frequency, and so on; with computerized data, it is easy to perform innumerable subgroup analyses.

Clearly, this is hazardous: if each of these tests uses an alpha of .05, then multiple tests are highly prone to at least one type I error (erroneously rejecting the null hypothesis that the medication is ineffective). Even if the null hypothesis is actually true for every test that is done, the probably of producing at least one type I error is .4 when 10 tests with an alpha of .05 are done; with 20 tests, the risk is .64. Especially when the commercial or academic stakes are high, it is tempting to perform multiple post hoc subgroup analyses and then to report only those that are statistically significant, leaving the reader unable to determine whether these are merely the false-positive "tips of the iceberg" of a larger number of tests that did not attain significance.

This risk of type I errors is true whether the multiple subgroup analyses are prespecified or post hoc; but when they are prespecified, the researchers at least had a preexisting rationale for the hypotheses. In an effort to establish exactly what hypotheses will be tested prior to data collection, and to reduce selective reporting of only favorable results, the International Committee of Medical Journal Editors requires that studies are preregistered prior to data collection, and that this registration includes a statement of the study hypothesis and primary and secondary outcome measures. Some memorable post hoc subgroup analysis findings that have been criticized as probable type I errors (Sleight, 2000) include

- a report that diabetes is commoner among boys born in October than in other months;
- a report that lung cancer is more common among people born in March than in other months;
- an analysis of the Second International Study of Infarct Survival (ISIS-2), which showed that aspirin had a markedly beneficial effect on heart attack patients, unless the patient happened to be born under the Gemini or Libra astrological birth signs.

Subgroup analyses are also at risk for type II errors, as the sample sizes in each of the subgroups may not be large enough for the test to have sufficient power to reject a false null hypothesis. In general, subgroup analyses should be considered only as starting points for further research, rather than conclusive in their own right, especially when they are based on post hoc analyses.

Subgroup effects are sometimes said to be the result of **effect modification** or **interaction**. For example, a randomized controlled trial of a new drug may find that it has no effect on the risk of diabetes, but a subgroup analysis of the data may show that it apparently increases the risk of diabetes among men, but reduces it among women. In this case, gender is an **effect modifier**—a factor that influences the effect of the phenomenon (such as risk of diabetes) that is being studied, but it does so differently in different subgroups (men vs. women) in a way that can only be seen after the data have been obtained.

In this situation, the division of the participants into men and women may be termed **stratification** (this post hoc stratification, or **stratified analysis**, differs from the stratification, sometimes called prestratification, which may be done during the design phase of the study to prevent confounding, and is discussed in Chapter 5). Effect modifiers are felt to be real biological effects, which we want to understand, unlike confounders, which are artifacts of poor study design, and which we want to eliminate.

This example also demonstrates the concept of **interaction**—the interdependent operation of two or more biological causes to produce, prevent, or control an effect. In this case, the drug and the patient's gender interacted to produce an effect on the risk of diabetes (the whole was different from the sum of the parts); this would be called a "drug $\times$ gender interaction."

Nonparametric and Distribution-Free Tests

The previous sections have dealt with testing hypotheses about means, using t- and z-tests. These tests share three common features:

- Their hypotheses refer to a *population parameter*: the population mean. For this reason, such tests are called **parametric** tests.
- Their hypotheses concern *interval* or *ratio* scale data, such as weight, blood pressure, IQ, per capita income, measures of clinical improvement, and so on.
- They make certain assumptions about the distribution of the data of interest in the population—principally, that the population data are normally distributed. (As was shown earlier, the central limit theorem allows this assumption to be made, even when little is known about the population distribution, provided that random samples of sufficient size are used.)

There are other statistical techniques that do not share these features:

- They do not test hypotheses concerning parameters, so they are known as **nonparametric tests**.
- They do not assume that the population is normally distributed, so they are also called **distribution-free tests**.
- They are used to test nominal or ordinal scale data.

Such tests, however, have the disadvantage that they are generally *less powerful* than parametric tests.

CHI-SQUARE

The most important nonparametric test is the **chi-square** (χ^2) test, which is used for testing hypotheses about *nominal scale* data.

Chi-square is basically a test of *proportions*, telling us whether the proportions of observations falling in different categories differ significantly from those that would be expected by chance.

> For example, in tossing a coin 100 times, we would expect about 50% (or 50) of the tosses to fall in the category of heads and 50% to fall in the category of tails. If the result is 59 heads and 41 tails, chi-square would show whether this difference in proportion is too large to be expected by chance (i.e., whether it is statistically significant).

As with other tests, chi-square involves calculating the test statistic (χ^2_{calc}) according to a standard formula and comparing it with the critical value (appropriate for the level of α selected) shown in the published chi-square tables, which can be found in most statistics textbooks.

Chi-square is also used in more complicated nominal scale questions. For example, a study might compare the rates of microbiologic cure of three different antibiotics used for urinary tract infections, as shown in Table 3-2. This kind of table is a **contingency table**, which is the usual way of presenting this kind of data. It expresses the idea that one variable (such as cure vs. lack of cure) may be contingent on the other variable (such as which antibiotic a patient took). The question that chi-square can answer is this: Is there a relationship between which antibiotic the patient took and achieving microbiologic cure?

TABLE 3-2	NUMBER OF PATIENTS ACHIEVING MICROBIOLOGIC CURE BY ANTIBIOTIC		
	Antibiotic A	Antibiotic B	Antibiotic C
Number achieving cure	49	112	26	Total 187
Number not achieving cure	12	37	8	Total 57
Total	61	149	34	

Chapter 4

Correlational and Predictive Techniques

Biomedical research often seeks to establish whether there is a relationship between two or more variables; for example, is there a relationship between salt intake and blood pressure, or between cigarette smoking and life expectancy? The methods used to do this are **correlational** techniques, which focus on the "co-relatedness" of the two variables. There are two basic kinds of correlational techniques:

1. **Correlation**, which is used to establish and quantify the *strength* and *direction* of the relationship between two variables.
2. **Regression**, which is used to express the *functional relationship* between two variables so that the value of one variable can be *predicted* from knowledge of the other.

Correlation

Correlation simply expresses the strength and direction of the relationship between two variables in terms of a **correlation coefficient**, signified by r. Values of r vary from -1 to $+1$; the strength of the relationship is indicated by the size of the coefficient, while its direction is indicated by the sign.

A plus sign means that there is a **positive correlation** between the two variables—high values of one variable (such as salt intake) are associated with high values of the other variable (such as blood pressure). A minus sign means that there is a **negative correlation** between the two variables—high values of one variable (such as cigarette consumption) are associated with low values of the other (such as life expectancy).

If there is a "perfect" linear relationship between the two variables so that it is possible to know the exact value of one variable from knowledge of the other variable, the correlation coefficient r will be exactly ±1.00. If there is absolutely no relationship between the two variables, so that it is impossible to know anything about one variable on the basis of knowledge of the other variable, then the coefficient will be zero. Coefficients beyond ±0.5 are typically regarded as strong, whereas coefficients between 0 and ±0.5 are usually regarded as weak. USMLE does not require the actual calculation of r.

SCATTERGRAMS AND BIVARIATE DISTRIBUTIONS

The relationship between two correlated variables forms a **bivariate distribution**, which is commonly presented graphically in the form of a **scattergram**. The first variable (salt intake, cigarette consumption) is usually plotted on the horizontal (X) axis, and the second variable (blood pressure, life expectancy) is plotted on the vertical (Y) axis. Each data point represents one observation of a pair of values, such as one patient's salt intake and blood pressure, so the number of plotted points is equal to the sample size n. Figure 4-1 shows four different scattergrams.

Determining a correlation coefficient involves mathematically finding the "line of best fit" to the plotted data points. The relationship between the appearance of the scattergram and the correlation coefficient can therefore be understood by imagining how well a straight line could fit the plotted points. In Figure 4-1A, for example, it is not possible to draw any straight line that

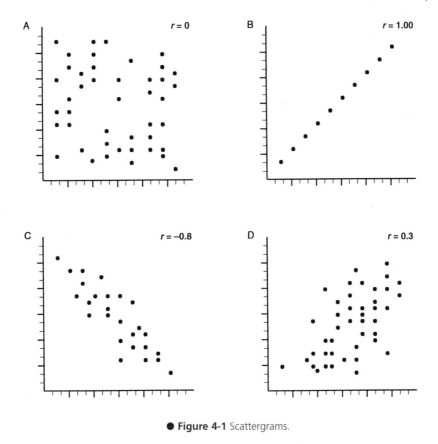

● **Figure 4-1** Scattergrams.

would fit the plotted points at all; therefore, the correlation coefficient is approximately zero. In Figure 4-1B, a straight line would fit the plotted points perfectly—so the correlation coefficient is 1.00. Figure 4-1C shows a strong negative correlation, with a correlation coefficient in the region of −0.8, and Figure 4-1D shows a weak positive correlation of about +0.3.

TYPES OF CORRELATION COEFFICIENT

The two most commonly used correlation coefficients are the **Pearson product-moment correlation**, which is used for *interval* or *ratio* scale data, and the **Spearman rank-order correlation**, which is used for *ordinal* scale data. The latter is sometimes symbolized by the letter ρ (rho). Pearson's r would be used, for example, to express the association between salt intake and blood pressure (both of which are ratio scale data), while Spearman's ρ would be used to express the association between birth order and class position at school (both of which are ordinal scale data).

Both these correlational techniques are **linear**: they evaluate the strength of a "straight line" relationship between two variables; if there is a very strong **nonlinear** relationship between two variables, the Pearson or Spearman correlation coefficients will be an underestimate of the true strength of the relationship.

Figure 4-2 illustrates such a situation. A drug has a strong effect at medium dosage levels, but very weak effects at very high or very low doses. Because the relationship between dose and effect is so nonlinear, the Pearson r correlation coefficient is low, even though there is actually a very strong relationship between the two variables. Visual inspection of scattergrams is therefore invaluable in

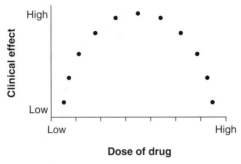

Figure 4-2 A strong nonlinear relationship.

identifying relationships of this sort. More advanced nonlinear correlational techniques can be used to quantify correlations of this kind.

COEFFICIENT OF DETERMINATION

The coefficient of determination expresses the proportion of the variance in one variable that is accounted for, or "explained," by the variance in the other variable. It is found by *squaring the value of r*, and its symbol is therefore r^2. So if a study finds a correlation (r) of 0.40 between salt intake and blood pressure, it could be concluded that $0.40 \times 0.40 = 0.16$, or 16% of the variance in blood pressure in this study is accounted for by variance in salt intake.

A correlation between two variables does not *demonstrate a causal relationship between the two variables,* no matter how strong it is. *Correlation is merely a measure of the variables' statistical association, not of their causal relationship—so the correlation between salt intake and blood pressure does not necessarily mean that the changes in salt intake* caused *the changes in blood pressure.* Inferring a causal relationship between two variables on the basis of a correlation is a common and fundamental error.

Furthermore, the fact that a correlation is present between two variables in a sample does not necessarily mean that the correlation actually exists in the population. When a correlation has been found between two variables in a sample, the researcher will normally wish to test the null hypothesis that there is no correlation between the two variables (i.e., that $r = 0$) in the population. This is done with a special form of t-test.

Regression

If two variables are highly correlated, it is possible to *predict* the value of one of them (the dependent variable) from the value of the other (the independent variable) by using **regression** techniques. In **simple linear regression**, the value of one variable (X) is used to predict the value of the other variable (Y) by means of a simple linear mathematical function, the **regression equation**, which quantifies the straight-line relationship between the two variables. This straight line, or **regression line**, is actually the same "line of best fit" to the scattergram as that used in calculating the correlation coefficient.

The simple linear regression equation is the same as the equation for any straight line:

$$\text{Expected value of } Y = a + bX,$$

where

> a is a constant, known as the "intercept constant" because it is the point where the Y axis is intercepted by the regression line (in other words, the value of Y when X is zero).

b is the slope of the regression line and is known as the **regression coefficient**; it shows the change in Y when X increases by 1 unit.

X is the value of the variable X.

Once the values of a and b have been established, the expected value of Y can be predicted for any given value of X. For example, it has been shown that the hepatic clearance rate of lidocaine (Y, in mL/min/kg) can be predicted from the hepatic clearance rate of indocyanine green dye (X, in mL/min/kg), according to the equation $Y = 0.30 + 1.07X$, thus enabling anesthesiologists to reduce the risk of lidocaine overdosage by testing clearance of the dye (Zito & Reid, 1978).

MULTIPLE REGRESSION

Other techniques generate **multiple regression** equations, in which more than one variable is used to predict the expected value of Y, thus increasing the overall percentage of variance in Y that can be accounted for; a multiple regression equation is therefore:

$$\text{Expected value of } Y = a + b_1 X_1 + b_2 X_{2+} \ldots b_n X_n$$

For example, Angulo *et al.* (2007) found that the risk of hepatic fibrosis (Y, the patient's fibrosis score) in patients with nonalcoholic fatty liver disease (NAFLD) could be predicted on the basis of the patient's age, body mass index (BMI), presence of diabetes or impaired fasting glucose (IFG), aspartate aminotransferase/alanine aminotransferase (AST/ALT) ratio, platelet count, and albumin level according to the multiple regression equation

$$Y = -1.675 + 0.037 \times \text{age (years)} + 0.094 \times \text{BMI (kg/m}^2) + 1.13 \times \text{presence of IFG}$$
$$\text{or diabetes (yes} = 1, \text{no} = 0) + 0.99 \times \text{AST/ALT ratio} - 0.013 \times \text{platelet count}$$
$$(\times 10^9\text{/L}) - 0.66 \times \text{albumin (g/dL)}$$

Use of this regression equation was shown to allow a liver biopsy to be avoided in the majority of patients of NAFLD, a disorder that is found in a substantial proportion of Western populations.

Multiple regression can also be used to search for other, potentially confounding, contributions to the condition of interest. If data on multiple other variables are collected, regression techniques would allow their contribution to the left side of the equation to be included. If, in doing so, the coefficients of the variables of interest change, then these variables have automatically been adjusted for the presence of the new (and apparently confounding) factors. If the coefficients do not change, then the new factors have no influence and are not confounders; if the coefficient of any variable approaches zero, it clearly plays no role in the outcome of interest and can therefore be removed from the equation.

LOGISTIC REGRESSION

In linear regression, Y can theoretically have any possible value. In many situations, however, we are not interested in a numerical outcome (such as the risk of fibrosis), but in a nominal or categorical one (typically a binary or dichotomous one, such as death vs. survival, developing lung cancer vs. not developing lung cancer).

Here, the relationship between risk factors and the outcome is clearly not a linear one: while exposure to a risk factor increases gradually, the change in outcome from survival to death, or from health to a diagnosis of cancer, is a sudden, all-or-nothing one. Hence, the outcomes (such as death vs. survival) are conventionally represented by 0 and 1.

A mathematical function, the **logistic function**, is used to transform the linear regression data so that the values of Y are limited to the range of 0 to 1, giving us the probability of the outcome Y occurring (like any probability, it can range only from 0 to 1) for given values of X. Just as with multiple regression, coefficients show the strength of the influence of different factors; some factors may have a very low coefficient, showing that they are not significant predictors of the outcome.

For reasons beyond the scope of USMLE, logistic regression is particularly suited to the analysis of case–control studies (to be discussed in Chapter 5), telling us the odds of the outcome of interest (such as death) versus its converse (such as recovery) for a given degree of exposure to various risk factors, as well as the *p* value associated with this finding. For this reason, logistic regression is used very commonly, particularly in public health and cancer research.

For example, D'Souza *et al.* (2007) compared patients (cases) with newly diagnosed oropharyngeal cancer with otherwise similar clinic patients (controls) who did not have cancer. A logistic regression analysis showed that multiple risk factors (HPV-16 seropositivity, poor dentition, infrequent toothbrushing, family history of squamous-cell carcinomas of the head and neck, and heavy tobacco use) were significantly associated with the development of this cancer, accounting for the development of 90% of the cases.

Survival Analysis

A further technique for analyzing data about binomial outcomes, such as death versus survival, or cancer versus no cancer, is provided by **survival analysis**. Rather than merely looking at the proportion of patients who die or survive, or at the likelihood of death or survival following exposure to various risk factors or treatments, survival analysis addresses questions to do with "time to event" data, such as

- What is the survival time of patients with a given disease?
- How does this survival time compare between two groups of patients given different treatments for the same disease?
- Among those who survive up to a certain time, at what time or rate can they be expected to die?
- What factors, such as risk factors or treatments, affect the survival time?

Despite its name, survival analysis can be used to help answer questions of this kind about any kind of event, not just death—such as how long it takes before a hip replacement fails, how long it takes before a patient has a second heart attack, or how long it takes for coronary stents to become restenosed, and it is widely used outside medicine, in fields such as engineering (how long is a hard drive or light bulb likely to survive?).

For example, a study of the survival times of patients with newly diagnosed prostate cancer might aim to discover the following:

- How long do patients with treated and untreated prostate cancer survive after diagnosis?
- What variables (risk factors, treatments) affect survival time, and to what extent do changes in each variable affect survival time?

Answering the first question appears simple. However, the mean or median survival time of treated and untreated prostate cancer patients cannot be determined for a number of reasons, the most obvious of which is that data from patients who are still alive at the end of the study is excluded (unless we continue to gather data until all patients are dead, which is not usually practical). Furthermore, some patients may be lost to follow-up, and others will die not of prostate cancer, but of unrelated causes. Omitting data from these patients would clearly bias the estimate of survival time. Such items of data are called **censored observations**, meaning that the event of interest had not occurred (or the data was not available) by the time the data was gathered.

The simplest technique in survival analysis is **life table analysis**. A life table tracks "lives" for the event of interest (such as death) in a group of people from a time when none of them have experienced the event until the end of the study, when data is no longer being collected. To analyze it, we divide the period under study into a certain number of equal intervals, such as 60 monthly intervals. For each of these intervals, we record the number of people who entered that time period alive, and the number who died during that time period, as well as the number of people who were lost to follow-up (censored) during that same period.

This allows us to calculate the **survival function**, which is the cumulative proportion of patients surviving at any particular time t (in other words, the probability of surviving beyond that time):

$$\text{Survival function at time } t = \frac{\text{Number of patients surviving longer than } t}{\text{Total number of patients}}$$

This calculation is called a **Kaplan–Meier analysis**; the survival function is recalculated each time an event (such as death) occurs.

The median survival of a group of people is the time when the survival function is 0.5. This is the same as the 50th percentile for survival time; in a similar way, 25th, 75th, or other percentiles can be found. (Note that the survival function solves the problem of censored observations, as it does not require knowing the survival time of every member of the group. Note also that the median or 50th percentile survival time will not be the point at which 50% of the patients have survived, unless there had been no censored observations until this time.)

Life table analysis also allows us to estimate the likelihood of further survival at any time (e.g., if a patient has survived for 3 years since diagnosis of prostate cancer, how likely is he to survive for a further 3 years?).

Life table data and the survival function are commonly shown graphically by means of a **Kaplan–Meier plot**, which is the commonest type of **survival curve** (although it is actually a series of steps, rather than a curve, as the proportion of people surviving changes discretely when the event of interest occurs). This plot shows the proportion of people surviving for any given length of time. Often there will be two or more groups of patients (such as those who were treated and those who were not treated, or those given treatment A vs. treatment B), and so there two or more curves. This is seen in Figure 4-3, which shows the proportions of under 65-year-old patients with

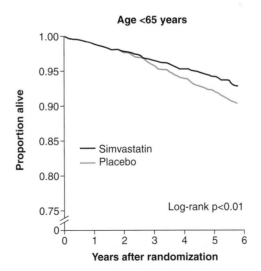

● **Figure 4-3** Kaplan–Meier plot. (Reproduced from Miettinen TA, Pyorala K, Olsson AG, *et al.* Cholesterol-lowering therapy in women and elderly patients with myocardial infarction or angina pectoris: findings from the Scandinavian Simvastatin Survival Study (4S). *Circulation.* 1997;96:4211–4218.)

established coronary heart disease who survived for the first 6 years after randomization to a statin drug (simvastatin) versus a placebo in the well-known 4S study (Scandinavian Simvastatin Survival Study).

The effect of a factor such as treatment is shown by the difference between the two curves. If the curves overlap or are very close, the difference between the treatments is unlikely to achieve statistical significance; specialized forms of the chi-square test (such as the **log rank** test, also known as the **Mantel–Haenszel** test) are used to test the null hypothesis that there is no difference between the two survival curves in the population. Note also that the distinction between the two curves is a nominal or categorical one—treatment versus placebo, or treatment A versus treatment B.

The second question in our prostate cancer study (e.g., "What variables, such as risk factors or treatments, affect survival time, and to what extent do changes in each variable affect survival time?") cannot be answered by a life table or Kaplan–Meier analysis, as it goes beyond asking about the effects of a nominal scale variable (treatment vs. no treatment, treatment A vs. treatment B).

This second question is similar to the questions that are answered by multiple regression and logistic regression techniques, in which a series of explanatory variables (X_1, X_2, X_3, etc.) are used to predict the value of variable Y. However, neither multivariate regression nor logistic regression can be used to answer this question because

- neither of them can deal with the problem of censored observations, and
- logistic regression can only predict the probability of death (the outcome Y), but cannot predict the likelihood of survival to a particular time, or the extent to which changes in each variable affect the likelihood of survival to a particular time.

A special regression technique, **Cox regression**, overcomes these problems. Like the other regression techniques, it produces an equation for the outcome Y (likelihood of survival to a particular time) as a function of several explanatory variables (X_1, X_2, X_3, etc.) (such as treatment, patient age, tumor stage etc.) simultaneously, and it allows us to see the strength and direction of the relative contributions of these variables, in the form of the coefficient for each variable.

This technique is also known as **Cox proportional hazards analysis**, because it shows the degree to which changes in the independent variables affect **hazard** (also called **hazard rate**)—the risk that an event (such as death) will occur at a given time, given that it has not occurred up until that time.

By looking at the hazard rates for different groups of patients (such as treated vs. untreated patients), Cox regression can produce an estimate of the **hazard ratio**: this is the ratio of the risk of the event at a particular time (the hazard) in the exposed group to the hazard in the unexposed group. For example, the hazard ratio for death from treated versus untreated prostate cancer 5 years after diagnosis might be 0.5, meaning that the treated patients are only half as likely to die at that time. Cox regression is therefore commonly used in clinical trials and in cohort studies.

A typical use of Cox regression was in the well-known Seven Countries Study (Jacobs et al., 1999) in which a cohort of over 12,000 middle-aged European, American, and Japanese men were followed for 25 years. Their smoking habits, country of residence, age, BMI, serum cholesterol, systolic blood pressure, and clinical cardiovascular disease were recorded.

After adjusting for each of the other variables, the hazard ratio for death (due to any cause) in those who smoked less than 10 cigarettes per day (vs. never-smokers) was 1.3; it was 1.8 for those who smoked more than 10 cigarettes per day. Hazard ratios were also elevated among smokers versus never-smokers for death due to (among other things) coronary heart disease, stroke, and lung and other cancers. *P* values were reported for each of these ratios, showing that they were unlikely to have been due to chance; hence this study confirmed that smoking is an independent risk factor for death due to these diseases.

Choosing an Appropriate Inferential or Correlational Technique

The basic choice of statistical technique for analyzing a particular set of data is determined by two factors: the scale of measurement and the type of question being asked. USMLE will require familiarity with only those techniques that have been covered here (although there are many others). Table 4-1 summarizes the choices:

Concerning *nominal scale data,* two kinds of question have been discussed:

- "Do the proportions of observations falling in different categories differ significantly from the proportions that would be expected by chance?" The technique for such questions is the **chi-square test**.
- Questions concerning prediction: "What is the likely outcome (such as life or death, sickness or wellness) of exposure to given risk factors?" The technique for such questions is **logistic regression**.

Regarding *ordinal scale data,* only one kind of question has been covered: "Is there an association between ordinal position on one ranking and ordinal position on another ranking?" The appropriate technique here is the **Spearman rank-order correlation**.

For *interval* or *ratio scale data,* two general kinds of questions have been discussed:

- Questions concerning means, such as "What is the true mean of the population?" or "Is one sample mean significantly different from another sample mean?" To answer this, a *t*-test will normally be used. Alternatively, provided that $n > 100$, or if the standard deviation of the population is known, a z-test may be used with virtually identical results.
- Questions concerning association and prediction, such as
  - "To what degree are two variables correlated?" To evaluate the strength and direction of the relationship, **Pearson product-moment correlation** is used, together with a form of *t*-test to test the null hypothesis that the relationship does not exist in the population.

TABLE 4-1

QUESTIONS CONCERNING	SCALE OF DATA		
	Nominal	Ordinal	Interval or Ratio
Differences in proportion	χ^2		
One or two means			*t*-test (or *z*-test if $n > 100$)
Association		Spearman ρ	Pearson *r*
Predicting the value of one variable: on the basis of one other variable			Simple linear regression
on the basis of multiple variables	Logistic regression (predictor variables are ratio data; predicted variable is nominal)		Multiple regression
Predicting survival time			Cox regression

- "What is the predicted value of one variable, given values of one or more other variables?" To make predictions about the value of one variable on the basis of one other variable, **simple linear regression** is used; if the prediction is to be based on several variables, then **multiple regression** is used; if the prediction is about survival time to an event, **Cox regression** is used.

Table 4-1 summarizes the range of inferential and correlational techniques that have been covered. This table should be memorized to answer typical USMLE questions that require choosing the correct test or technique for a given research situation.

Chapter 5

Asking Clinical Questions: Research Methods

Medical research typically aims to discover the relationship between one or more events or characteristics (such as being exposed to a toxic substance, having a family history of a certain disease, or taking a certain drug) and others (such as contracting or overcoming an illness). All these events or characteristics are called **variables**.

Independent variables are presumed to be the *causes* of changes in **dependent variables**, which are presumed to *depend* on the values of the independent variables. Research therefore typically seeks to uncover the relationship between independent and dependent variables. For example, in studying the effectiveness of a particular drug for a disease, the use or nonuse of the drug is the independent variable, and the resulting severity of the disease is the dependent variable.

Unless a research project studies an entire population, it will involve the use of a sample—a subset of the population. Most samples used in biomedical research are **probability samples**—samples in which the researcher can specify the probability of any one element in the population being included. For example, if a sample of 1 playing card is picked at random from a pack of 52 cards, the probability that any 1 card will be included is 1/52. Probability samples permit the use of inferential statistics, while nonprobability samples allow the use of only descriptive statistics.

A sample is **representative** if it closely resembles the population from which it is drawn. *All types of random samples tend to be representative, but they cannot guarantee representativeness.* Nonrepresentative samples can cause serious problems.

A classic example of a nonrepresentative sample was an opinion poll taken before the 1936 US Presidential Election. On the basis of a sample of more than 2 million people, it was predicted that Alfred Landon would achieve a landslide victory over Franklin Delano Roosevelt, but the result was the opposite. The problem? The sample was drawn from records of subscribers to a literary magazine and telephone and automobile ownership—and such people in that Depression year were not at all representative of the electorate as a whole.

A sample or a result demonstrates **bias** if it **consistently errs in a particular direction**. For example, in drawing a sample of 10 from a population consisting of 500 white people and 500 black people, a sampling method that consistently produces more than 5 white people would be biased. *Biased samples are therefore unrepresentative, and true randomization is proof against bias.*

This kind of bias is called **selection bias** or **sampling bias**. Bias from nonrepresentativeness may be due to simple errors in drawing the sample, as in the 1936 opinion poll, or due to more subtle influences: Patients with a certain disease who are referred to a specialist or to a teaching hospital are likely to have more severe disease than other patients in the community with the same disease (this is **referral bias**); patients without insurance may not be diagnosed or come to clinical attention at all. **Self-selection** may also occur: Certain kinds of people may volunteer to take part in a study, or be attracted by payment for participating or by the promise of free care, and certain kinds may refuse to participate for various reasons; this results in **participation bias** or **volunteer bias**.

There are four basic kinds of probability samples: simple random samples, stratified random samples, cluster samples, and systematic samples.

Simple Random Samples

This is the simplest kind of probability sample—a sample drawn in such a way that every element in the population has an equal probability of being included, such as in the playing card example above. A random sample is defined by the *method of drawing the sample, not by the outcome*. If four hearts were picked in succession out of the pack of cards, this does not in itself mean that the sample is not random (although the sample would clearly not be representative); if the sampling method frequently tended to produce four hearts, it would clearly be biased and not truly random.

Stratified Random Samples

In a stratified random sample, the population is first divided into internally relatively homogeneous groups, or **strata**, from which random samples are then drawn. This stratification results in greater representativeness. For example, instead of drawing one sample of 1,000 people from a total population consisting of 1 million men and 1 million women, one random sample of 500 could be taken from each gender (or stratum) separately, thus guaranteeing the gender representativeness of the resulting overall sample of 1,000 and eliminating selection bias as far as the gender of the participants go. If 500 women could not be found to participate in the study, but only (say) 250, it would be possible to correct for the lack of women by multiplying (or **weighting**) data obtained from the women in the sample by 2.

Cluster Samples

Cluster samples may be used when it is too expensive or laborious to draw a simple random or stratified random sample. For example, in a survey of 100 US medical students, an investigator might start by selecting a random set of groups or "clusters"—such as a random set of 10 US medical schools—and then interviewing 10 students in each of those 10 schools. This method is much more economical and practical than trying to take a random sample of 100 directly from the population of all US medical students.

Systematic Samples

These involve selecting elements in a systematic way—such as every fifth patient admitted to a hospital or every third baby born in a given area. This type of sampling usually provides the equivalent of a simple random sample without actually using randomization.

Experimental Studies

The relationship between independent and dependent variables can be investigated in two ways:

- **Experimental** studies, in which the researcher exercises control over the independent variables by deliberately manipulating them; experimental studies are sometimes called **intervention** studies.

- **Nonexperimental** studies, in which nature is simply allowed to take its course; nonexperimental studies are also called **observational** studies.

> In an *experimental* investigation of a drug's effectiveness, for example, the investigator would intervene, giving the drug to one group of patients, but not to another group. In a *nonexperimental* investigation, the researcher would simply observe different patients who had or had not taken the drug in the normal course of events. The hallmark of the experimental method is therefore manipulation or intervention.

Experimental studies of new treatments follow a series of stages. In the United States, these stages are regulated by the U.S. Food and Drug Administration (FDA); preclinical animal studies are reviewed by the FDA before an Investigational New Drug (IND) application is approved and clinical trials can begin. Clinical trials fall into three initial phases:

- **Phase 1** studies generally involve a small number (under 100) of healthy volunteers, and the dependent variables are typically the safety, side effects, and metabolic and excretory pathways of the drug.
- If Phase 1 studies show no unacceptable toxicity, **Phase 2** studies are done on somewhat larger numbers of people who actually have the condition being treated. These studies usually have a control group (see below), and the outcomes studied are the clinical effects on the condition and the treatment's short-term side effects and safety.
- If the drug seems to be effective, **Phase 3** studies are done: controlled trials with larger numbers of patients (typically several hundred or a few thousand) at a range of geographical sites (**multicenter** studies), focusing on efficacy and safety. They often use a range of doses of the drug in different populations of patients. Depending on the findings of Phase 3 trials, the developer of the drug may submit a New Drug Application (NDA); the FDA then reviews all the data, confirms that the facilities in which the drug is going to be manufactured meet certain standards, and may then approve or deny the application, request further studies, or request the opinion of an independent advisory committee.

CLINICAL TRIALS

The experimental method in medical research commonly takes the form of the **clinical trial**, typically a randomized controlled clinical trial, which attempts to evaluate the effects of a treatment. Clinical trials are the "gold standard" of evidence for cause-and-effect relationships. However, they have disadvantages:

- They may pose multiple ethical problems (discussed below).
- They are impractical if the cause-and-effect relationship is one that takes a long time to appear.
- They are very expensive and time-consuming.
- They require that a treatment is adequately standardized or developed (e.g., treatments such as diets, exercise, or psychotherapy may be hard to fully standardize, making the results hard to interpret).

CONTROL GROUPS

Controlled clinical trials involve dividing participants or patients into two groups:

1. The **experimental** group, which is given the treatment under investigation.
2. The **control** group, which is treated in exactly the same way as the experimental group except that it is not given the treatment.

Any difference that appears between the two groups at the end of the study can then be attributed to the treatment under investigation. Control groups therefore help to eliminate alternative explanations for a study's results.

> For example, if a drug eliminates all symptoms of an illness in a group of patients in 1 month, it may be that the symptoms would have disappeared spontaneously over this time even if the drug had not been used. But if a similar control group of patients who did not receive the drug experienced no improvement in their symptoms, this alternative explanation is untenable.

There are two main types of control groups used in medical research:

1. A **no-treatment** control group, which receives no treatment at all. This leaves open the possibility that patients whose symptoms were relieved by the drug were not responding to the specific pharmacologic properties of the drug, but to the nonspecific placebo effect that is part of any treatment.

2. A **placebo** control group, which is given an inert placebo treatment. This allows us to eliminate the explanation that patients in the treatment group were responding to the placebic component of the treatment. The effectiveness of the drug would therefore have to be attributed to its pharmacologic properties.

In studies of this kind, it is obviously important that patients do not know whether they are receiving the real drug or the placebo: if patients taking the placebo knew that they were not receiving the real drug, the placebo effect would probably be greatly reduced or eliminated. Likewise, the researchers administering the drug and assessing the patients' outcomes must not know which patients are taking the drug and which are taking the placebo. Knowledge or deduction of which group the patient is in, by either the patient or the researchers, may cause conscious or unconscious bias: This is **ascertainment bias** or **detection bias**. Specifically, ascertainment bias on the part of the researchers is called **assessor bias** and on the part of the patients is called **response bias**.

The patients and all those involved in the conduct of the experiment should therefore be "blind" as to which patients are in which group; such studies are therefore called **double-blind** studies. However, it is not always possible to perform a double-blind study.

> For example, in an experiment comparing a drug versus a surgical procedure, it is hard to keep patients "blind" as to which treatment they received (although studies have been done in which control group patients have undergone sham surgery under general anesthesia). However, it would be simple for the outcome to be measured by a "blind" rater, who might perform laboratory tests or interviews with the patient without knowing to which group the patient belonged—such a study would be called a **single-blind** study.

> Under some circumstances, truly controlled experiments may not be possible. In research on the effectiveness of psychotherapy, for example, patients who are placed in a no-treatment control group may well receive help from friends, family, clergy, self-help books, and so on, and would therefore not constitute a true no-treatment control. In this case, the study would be called a **partially controlled** clinical trial.

Clinical trials aim to isolate one factor (such as use of a drug) and examine its contribution to patients' health by holding all other factors as constant as possible. Factors other than the experimental treatment that are not constant across the control and experimental groups are called **confounders**.

For example, in a controlled trial of a new drug (vs. a placebo) for the prevention of prostate cancer, a disease that is known to affect black men more than white men, it would be absurd to allocate all the white patients to the drug group, and all the black patients to the placebo group; this would result in **allocation bias**, and any difference in outcome between the two groups could be attributed to differences between the races rather than to the drug itself.

In this situation, patient race is a potential **confounding variable**: it is likely to contribute differently to the two groups, so that the drug and the placebo groups are not directly comparable with each other. Thus, it is impossible to attribute differences in outcome to the treatment. Note that any effect of race in this would be an artifact of study design; if race did actually have a true biological effect (as an effect modifier, as discussed in Chapter 3), this poorly designed study would not be able to detect it.

Confounding is prevented or minimized by several techniques of study design, including **randomization**, **matching**, **stratification**, and **restriction**.

RANDOMIZATION

Randomization means that patients are randomly assigned to different groups (i.e., to the experimental and control groups). Most clinical trials involve randomization and are therefore called **randomized clinical trials** (RCTs), or **randomized controlled clinical trials** (RCCTs). True randomization means that the groups should be similar with respect to race, gender, disease severity, age, occupation, and any other variable (including variables that we do not even recognize) that may affect the response to the experimental intervention.

MATCHING

Randomization cannot *guarantee* that the experimental and control groups are similar in all important ways. An alternative way of ensuring similarity is by **matching**: each patient in the experimental group is paired with a patient in the control group who is a close match on all relevant characteristics—so if gender, race, age, and smoking status were important factors influencing the course of the disease being studied, each experimental patient would be matched with a control patient of the same gender, race, age, and smoking status. Thus, any resulting differences between the two groups could not be attributed to differences in these factors. However, it may be difficult or expensive to find a matching participant for each person in the trial, and variables that are not matched may still turn out to be confounders.

STRATIFIED RANDOMIZATION (ALSO CALLED PRESTRATIFICATION OR BLOCKING)

This is a combination of randomization and matching techniques. The population under study is first divided, or **stratified**, into subgroups ("blocks") that are internally homogeneous with respect to the important potential confounding factors (e.g., gender, race, age, disease severity). Equal numbers of patients from each subgroup are then randomly allocated to the experimental and control groups. The two groups are therefore similar, but their exact membership is still a result of randomization.

RESTRICTION

This involves restricting a study to participants with certain characteristics—if a study is limited to people of just one gender, race, age group, geographical area, disease severity, etc., then all these factors are removed as potential confounders.

　　Recruitment and retention diagrams show the flow of participants through a study. Figure 5-1 shows an example in the form of a standard for such diagrams from the CONSORT (Consolidated Standards of Reporting Trials) group, endorsed by hundreds of leading medical journals.

　　As Figure 5-1 shows, not all participants in a study will produce a complete set of analyzable data; some participants may not have fully received the treatment they were allocated to, some may discontinue it, and some may just be missing or **lost to follow-up**.

　　If these participants were equally distributed between the control and treatment groups (in a clinical trial) or were similar to the remaining participants (in nonexperimental trials), the only problem this might cause would be a loss of sample size with consequent loss of statistical power. But "lost" participants are obviously likely to be different: they may have died from the disease under study or been institutionalized (perhaps because of, or in spite of, the study treatment or the placebo) and died of comorbid illnesses; they may have dropped out due to side effects from, or dissatisfaction with, the study treatment; they may have had social or transportation problems that made it too hard for them to follow-up, and so on. Patients who are lost to follow-up are likely to have worse outcomes than those who remain, thus biasing the results of the study in a favorable direction.

　　One way of dealing with the problem of participants for whom a complete set of data cannot be obtained is by **intention to treat analysis**, in which the data from all participants is analyzed according to the group they were allocated to, regardless of whatever happened subsequently—in other words, it is analyzed according to *how it was intended to treat them*. This overcomes the problem of bias and resembles the real clinical world in which patients do not all take the medications they are prescribed, but it does not solve the problem of participants whose data are missing. This missing data are sometimes filled in or **imputed**, typically by the **last observation carried forward** (LOCF) method, in which the last measurement obtained for a given participant is simply carried forward as if it were the same at later dates.

　　For example, a patient who dropped out after 6 months of a 2-year trial of an Alzheimer disease drug might have his mental status exam scores at 6 months carried forward to the 2-year conclusion.

　　However, this may introduce bias in ways that are hard to disentangle; the patient's LOCF data may make the drug appear to be less effective than it really is if he has failed to get the full 2-year benefit of an effective drug, or it may make the drug appear more effective than it really is if the other patients on the drug showed a progressive cognitive decline as the study went on.

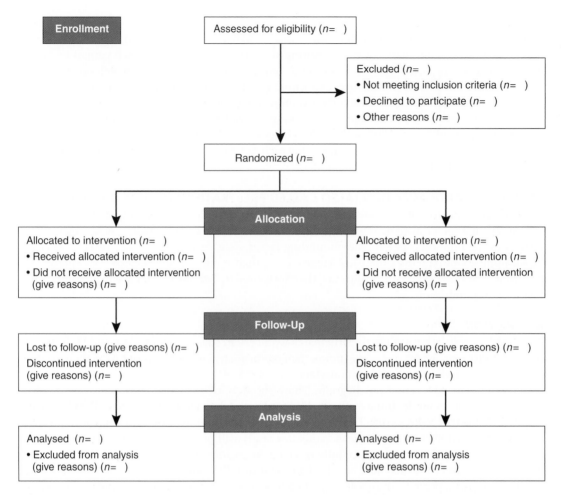

● **Figure 5-1** A template for a recruitment and retention diagram. (Reproduced with permission from Schulz KF, Altman DG, Moher D, for the CONSORT Group. CONSORT 2010 Statement: updated guidelines for reporting parallel group randomised trials. *BMJ.* 2010;340:c332.)

If a study is designed and performed well, including in its management of missing data, so that it is free of bias, and alternative explanations for its findings are ruled out, it has **internal validity**—it will be valid *for the particular participants it studied.* If the sample is truly representative of the population from which it was drawn, then the study has **external validity**—its findings can be applied *to that population.* However, it may not necessarily be **generalizable**—valid more broadly, to *other people, in other places, or at other times.*

All clinical trials will be limited to some extent, if only by the geographical area of the participants (such as an area within reach of the medical centers running the trial), and in most clinical research, participants will also be limited by certain **entry** or **inclusion criteria**, which must be clearly specified before the sample is drawn.

For example, in a trial of a new drug for Alzheimer disease, there may be criteria specifying the patients' ages, how the diagnosis was made, how long they have had the disease and how severe it is, the absence of certain comorbidities, and that they have not already taken any drug for the disease. The findings of such a study may not be generalizable to "real world" Alzheimer patients, who are likely to be much more varied, complex, and less well defined.

Entry and exclusion criteria have an ethical dimension if the sample studied is not representative of the population of people with the disease in question, or of the population as a whole. Until the 1980s, studies showed a notorious tendency to be restricted in favor of males and whites, but it is now a matter of public policy in the United States that proposed studies cannot restrict studies in these ways without strong justification.

Research Ethics and Safety

All research has an ethical dimension, and in the United States researchers are required to obtain permission from an **Institutional Review Board** (IRB) before collecting any human data, and must inform the IRB of any changes as the study progresses. The membership and procedures of IRBs are federally regulated. An IRB may be part of an academic institution, a hospital, or it may be a free-standing for-profit commercial entity.

IRB rules grew out of the scandal of the Tuskegee Syphilis Experiment, in which poor black men in rural Alabama who had been infected with syphilis were studied for several decades without being treated, some for many years even after penicillin was established as the standard and effective treatment in 1951. IRB rules require that

- participants' autonomy is respected;
- vulnerable people (e.g., elderly, seriously ill, prisoners) or those with diminished autonomy (e.g., children, mentally ill) are protected;
- benefits are maximized, and potential harms minimized;
- consent to participate is truly **informed consent**. The potential participant must be given sufficient information, in a way that he or she can understand (taking into account limitations of intelligence, rationality, maturity, language, etc.), and the consent must be voluntary, not the result of coercion or of excessive or inappropriate reward. Clearly, the amount of information given will vary with the type of study being done: the risk of a trial of new type of prosthetic heart valve is greater than the risk of just having blood drawn. Just as with consent to medical treatment, the "informed consent" form that participants sign is not the consent itself, but is only a record of the consent; it does not replace the process of obtaining informed consent.

IRBs are also required to ensure that standards are in place to ensure the privacy of research participants and maintain the confidentiality of data about them. IRBs, and the medical community as a whole, are paying increasing attention to financial and academic conflicts of interest among researchers and their sponsors, which potentially cause many unethical biases in the design, recruitment, conduct, interpretation, and reporting phases of clinical trials.

Ethical problems are particularly significant in experimental studies, especially if there is reason to believe that the independent variable under investigation is either a harmful one *or* a strongly beneficial one, even if these reasons only become apparent during the course of the study.

> For example, although it would be unethical to test the hypothesis that it takes 15 years of heavy alcohol drinking to cause cirrhosis of the liver by conducting an experiment, this hypothesis can be tested observationally by finding people who have done this in the ordinary course of events.

Similarly, if it is believed that an effective treatment for a condition already exists, the use of a placebo or no-treatment control group may be unethical; as a result, some long-standing time-honored medical interventions may be hard to study ethically (or legally) because they have become the standard of care, despite no real studies having been done to prove their efficacy.

> In the 1950s, experiments on the effects of oxygen on premature babies were opposed on the grounds that the control group would be deprived of a beneficial treatment; later, when it became strongly suspected that excessive oxygen was a cause of a type of blindness (retinopathy of prematurity), similar experiments were opposed because the experimental group might be subjected to a harmful treatment.

One solution is to assign the control group to a different active treatment; the new experimental treatment is then compared only with that treatment. This may be problematic for some research sponsors, such as pharmaceutical companies, who may wish to show that their new drug is more effective than a placebo, rather than subject it to a head-to-head contest with its competitors—especially as FDA approval rests on a drug being effective only versus a placebo, not on it being more effective than (or even as effective as) existing treatments.

Another solution, if the treatment is not time-sensitive, or if resources are inadequate to treat all patients at the same time, is to use a **wait list control group**: the control group patients will receive the treatment after the experimental group patients have received theirs; so the controls' treatment is merely delayed, not denied.

Related to this is the **crossover study**, in which half the patients receive the active treatment for a period, followed by the placebo, while the other half receive the placebo first, followed by the experimental treatment. This kind of study is called a **within-subjects** design: comparisons between drug and placebo are made *within* each subject (or patient), rather than *between* different subjects. In these designs, each patient serves as his or her own control (so the control group is a **same-subject** control group). This is also called a **repeated measures** study because the measurements of the dependent variable (such as symptom severity) are repeated within each patient, rather than a single measurement being taken, and compared between different patients in the more common **between-subjects** designs.

If there is a danger of a "carryover" effect (e.g., if the treatment is a drug that may continue to have some effect after it is withdrawn), there can be a **washout** period between the drug and the placebo phases, during which no treatment is given. Figure 5-2 illustrates a crossover design with washout: One group of patients receives the drug for 1 month and then "crosses over" to receive the placebo after a 1-month washout. The other group follows this pattern in reverse order.

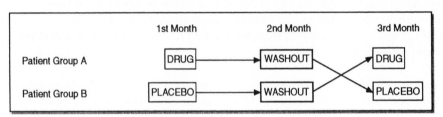

● **Figure 5-2** A crossover design with washout.

As more treatments become available for many diseases, and as it becomes harder to create truly novel treatments, **non-inferiority trials** are becoming common. This kind of clinical trial seeks to show only that a new treatment is *not inferior* to an existing one; this may be done for commercial reasons, or because the existing treatment may have significant problems (such as side effects, requiring lab monitoring, having to be given by injection or multiple times a day, etc.).

In non-inferiority trials, the null hypothesis is that the old treatment is more effective than the new one, and the alternative hypothesis is a **directional** one—that the new treatment is *at least as* effective as the old one. This is therefore a situation in which **one-tailed** statistical testing is appropriate (as discussed in Chapter 3) because there will be no practical difference between finding a difference that favors the new treatment and finding that there is no difference between the two treatments; in either case, the new treatment is simply at least as effective as the old one.

Major drug trials are subject to regular safety monitoring to check that participants are not being subjected to unexpected or excessive risks, and at certain points an **interim analysis** will be performed (e.g., after 1 year of a 3-year trial, or after 50% of participants have been enrolled) for several reasons:

- To check for risks for the treated patients (**safety monitoring**)
- To see whether the treatment has such significant benefits that it would be unethical (or unnecessary) to continue the study, depriving the control group patients of the treatment.
- To see whether there is such a small chance of the study producing a useful result that it is futile to continue it

Several trials have been discontinued for each of these reasons in recent years after the interim analyses were followed by the application of prespecified **stopping rules** or **stopping boundaries**. There is concern that in some cases, trials may be stopped prematurely, before truly robust conclusions can be drawn, for purely commercial reasons, such as to allow a new drug to be brought to market or to gain an approved indication more quickly.

A solution to potential ethical problems caused by randomizing participants to different treatments is to conduct a **patient preference trial**, in which patients who have preferences about their treatments are allocated to that treatment. A **patient preference arm** can be included in a regular randomized trial: patients who have a strong preference are given the treatment they want, while those who have no preference are randomized. This allows increased numbers of patients to be studied (e.g., if there are two different kinds of drugs for the disease under study, some patients may refuse to participate in a randomized trial, but may be willing to enroll in one in which they get the medication they prefer), and has the advantage of reflecting real clinical practice, where patients do express preferences about their treatments and are more likely to be motivated to comply when they are given the treatment they want.

Nonexperimental Studies

Nonexperimental (or observational) studies fall into two general classes: **descriptive studies** and **analytic studies**.

Descriptive studies aim to describe the occurrence and distribution of disease or other phenomena. They do not try to offer explanations or test a theory or a hypothesis, but merely attempt to generate a description of the frequency of the disease or other phenomenon of interest according to the places, times, and people involved. These studies will use descriptive statistics, but not inferential statistics.

Descriptive studies are often the first method used to study a particular disease—hence, they are also called exploratory studies—and they may serve to generate hypotheses for analytic studies to test. There are many well-known examples:

- The first finding that occupational or environmental exposures could cause cancer: in 1775 Sir Percival Pott, a British surgeon, described scrotal cancer commonly found among boys who were employed as chimney sweeps; before this, cancer was assumed to be the result of human biology only.
- A description of eight unusual cases of Kaposi sarcoma among gay males in California, which was the earliest study of AIDS. As with scrotal cancer, the true cause of the disease was unknown at the time, but the description generated useful hypotheses.

Analytic studies aim to test hypotheses or to provide explanations about a disease or other phenomena—hypotheses or explanations that are often drawn from earlier descriptive studies. They therefore use inferential statistics to test hypotheses.

Descriptive and analytic studies are not always entirely distinguishable. For example, a large-scale descriptive study may provide such clear data that it provides an answer to a question or gives clear support to a particular hypothesis.

NONEXPERIMENTAL DESIGNS

Descriptive or analytic studies use one of four principal research designs: They may be **cohort studies**, **case–control studies**, **case series studies**, or **prevalence surveys**.

Cohort Studies

Cohort studies focus on factors related to the development of a disease. A **cohort** (a group of people) that does *not* have the disease of interest is selected and then observed for an extended period. Some members of the cohort will already have been exposed to a suspected risk factor for the disease, and others will eventually become exposed; by following them all, the relationship between the risk factors and the eventual outcomes can be seen. This kind of study therefore allows the incidence and natural history of a disease to be studied. (Some cohort studies, however, do follow patients who do have the disease of interest; in particular, **inception cohorts** consist of patients who have just been diagnosed, and who are followed to determine their prognosis.)

Determining which people are classified as members of the "exposed" versus "unexposed" groups involves a decision called **exposure allocation**. For example, if we are interested in examining the relationship between heavy alcohol use and cirrhosis of the liver, we would need to define what constitutes "heavy" alcohol use and what does not.

Cohort studies may be loosely termed **follow-up** or **longitudinal** studies because they follow people over a prolonged period, tracing any changes through repeated observation. They are also called **prospective** studies because people are followed forward from a particular point in time, so the researcher is "prospecting" or looking for data about events that are yet to happen. Cohort studies are sometimes also called **incidence** studies because they look for the incidence of new cases of the disease over time.

> A famous example of a cohort study is the Framingham Study, which started in Framingham, Massachusetts in 1949 with a cohort of more than 5,000 people who were free of coronary heart disease (CHD). The individuals in the cohort were reexamined every 2 years for more than 30 years, and it allowed the first identification of the major physical risk factors for CHD.

Cohort studies have a number of significant advantages:

- *When a true experiment cannot be conducted (whether for ethical or practical reasons), cohort studies are the best form of investigation; their findings are often extremely valuable.*
- *They are the only method that can establish the absolute risk of contracting a disease, and they help to answer one of the most clinically relevant questions: If someone is exposed to a certain suspected risk factor, is that person more likely to contract the disease? Cohort studies may also reveal the existence of protective factors, such as exercise and diet.*

- *Because cohort studies are prospective, the assessment of risk factors in these studies is unbiased by the outcome.* If the Framingham Study were retrospective, for example, people's recollection of their diet and smoking habits could have been biased by the fact that they already have CHD (this effect is known as **recall bias**). In addition, the chronologic relationship between the risk factors and the disease is clear in prospective studies.
- For the individuals in a cohort who ultimately contract the disease of interest, data concerning their exposure to suspected risk factors has already been collected. However, in a retrospective study, this may not be possible. Again, if the Framingham Study were retrospective, it might have been impossible to obtain accurate information about the diet and smoking habits of people who had already died. Other types of studies are often unable to include people who die of the disease in question— often the most important people to study.
- Information about suspected risk factors collected in cohort studies can be used to examine the relationship between these risk factors and many diseases; therefore, a study designed as an analytic investigation of one disease may simultaneously serve as a valuable descriptive study of several other diseases.
- They allow the *chronological relationship* between exposure to the risk factor and development of the disease to be determined.

Cohort studies have some important disadvantages:

- *They are time-consuming, laborious, and expensive to conduct*: Members of the cohort must be followed for a long time (often for many years) before a sufficient number of them get the disease of interest. It may be very expensive and difficult to keep track of a large number of people for several years, and it may be many years before results are produced, especially in the case of diseases that take a long time to appear after exposure to a risk factor.

- *They may be impractical for rare diseases*: If one case of a disease occurs in every 10,000 people, then 100,000 people will have to be followed for 10 cases to eventually appear. However, if a particular cohort (such as an occupational cohort) exists with a high rate of the disease, a disease that is rare in the general population can still be studied by this method. A classic example was a study of workers in a vinyl chloride factory and their subsequent development of angiosarcoma of the liver.

Historical cohort studies. *Although cohort studies are traditionally prospective, there is second kind of cohort study: the* **historical** *(or* **retrospective***) cohort study, in which the researcher has access to information about the prior exposures of a particular group of people, and looks back at the records of all the people in this group to see whether exposure is related to disease.*

> A memorable if trivial example is a study of the "mummy's curse"—the notion that people who were associated with events surrounding the opening of King Tutankhamen's tomb in Egypt in the 1920s would die prematurely (Nelson, 2002). The author retrospectively reviewed data about the survival of two cohorts of Westerners: One cohort was simply present in Egypt at the time, while the other cohort was present with the archaeologist Howard Carter either when tomb, the sarcophagus, or the coffin were opened, or when the mummy was examined. No association was found between this exposure and premature death!

Case–Control Studies

While cohort studies examine people who are initially free of the disease of interest, **case–control** studies compare people who *do* have the disease (the cases) with otherwise similar people who *do not* have the disease (the controls).

Case–control studies start with the outcome, or dependent variable (the presence or absence of the disease). They then look *back* into the past for possible independent variables that may have caused the disease, to see whether a possible risk factor was present more frequently in cases than in controls; hence, they are also called **retrospective** studies.

> One classic exploratory case–control study uncovered the relationship between maternal exposure to diethylstilbestrol (DES) and carcinoma of the vagina in young women (Herbst *et al.*, 1971). Eight patients with this rare cancer were each compared with four matched cancer-free controls. Looking back at their individual and maternal histories, no significant differences appeared between the cases and the controls on a wide range of variables, but it was found that mothers of seven of the eight cases had taken DES in early pregnancy, 15 to 20 years earlier, while none of the 32 controls' mothers had done so.

Case–control studies are like historical cohort studies in that they both look back at the data (so they are "post hoc" studies), but they differ in that *all* the people initially identified in case–control studies have the disease.

Case–control studies offer some significant advantages:

- *They can be performed fairly quickly and cheaply (especially in comparison with cohort studies), even for rare diseases or diseases that take a long time to appear (as the vaginal carcinoma example shows). Because of this, case–control studies are the most important way of investigating rare diseases and are typically used in the early investigation of a disease.*

- They require comparatively few subjects.
- They allow multiple potential causes of a disease to be investigated.

Case–control studies also have a number of disadvantages and are particularly subject to bias:

- People's recall of their past behaviors or risk factor exposure may be biased by the fact that they now have the disease (recall bias).
- The only cases that can be investigated are those that have been identified and diagnosed; *undiagnosed or asymptomatic cases are missed.* People who have already died of the disease cannot be questioned about their past behaviors and exposure to risk factors.
- Selecting a comparable control group is a difficult task that relies entirely on the researcher's judgment.
- They cannot determine the rate or the risk of the disease in exposed and nonexposed people.
- They cannot prove a cause-and-effect relationship.

Case Series Studies

A **case series** simply describes the presentation of a disease in a number of patients. It does not follow the patients for a period, and it uses no control or comparison group. Therefore, it cannot establish a cause-and-effect relationship, and its validity is entirely a matter for the reader to decide—a report that 8 of a series of 10 patients with a certain disease have a history of exposure to a particular risk factor may be judged to be extremely useful or almost worthless.

Despite these serious shortcomings, case series studies are commonly used to present new information about patients with rare diseases, and they may stimulate new hypotheses; again, the original work on AIDS patients described just eight cases of Kaposi sarcoma. Case series studies can be done by almost any clinician who carefully observes and records patient data. A **case report** is a special form of case series in which only one patient is described—this too may be very valuable, or virtually worthless.

Prevalence Surveys

A **prevalence survey** or **community survey** is a survey of a whole population. It assesses the proportion of people with a certain disease (the **prevalence** of the disease; to be discussed in the next chapter) and examines the relationship between the disease and other characteristics of the population. It may also be used to find the prevalence of a disease in people who have and have not been exposed to a risk factor (e.g., chronic obstructive pulmonary disease in smokers vs. nonsmokers): this is the **prevalence ratio**.

Because prevalence surveys are based on a single examination of the population at a particular point in time and do not follow the population over time, they are also called **cross-sectional** studies, in distinction to longitudinal (cohort) studies.

Prevalence surveys are common in the medical literature—for example, a study could compare the prevalence of CHD in one community versus a different community with different dietary or exercise habits. Another example would be a study examining the prevalence of lung disease in one city, which could then be compared with another city with lower levels of cigarette consumption or air pollution.

Prevalence surveys suffer from a number of disadvantages:

- Because they look at existing cases of a disease, and not at the occurrence of new cases, they are likely to *overrepresent chronic diseases and underrepresent acute diseases.*
- *They may be unusable for acute diseases,* which few people happened to have at the moment they were surveyed.
- People with some types of disease may leave the community, or may be institutionalized, causing them to be excluded from the survey.

Findings of prevalence surveys must be interpreted cautiously; the mere fact that two variables (such as high fish intake and low prevalence of coronary disease) are associated does not mean that they are causally related.

Although they are expensive and laborious, prevalence surveys are common because they can produce valuable data about a wide range of diseases, behaviors, and characteristics. These data can be used to generate hypotheses for more analytic studies to examine.

Figure 5-3 summarizes the chronological sequence of events in the major kinds of nonexperimental study designs.

Cohort studies

EXPOSURE   STUDY STARTS   FURTHER EXPOSURE   DISEASE

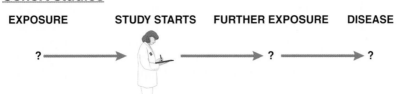

Historical cohort studies

EXPOSURE   DISEASE   STUDY STARTS

EXPOSURE   DISEASE   STUDY STARTS

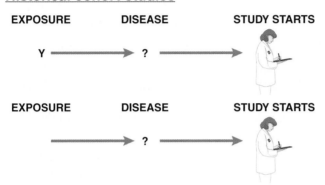

Case-control studies

EXPOSURE   DISEASE   STUDY STARTS

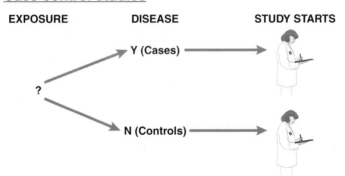

Case series studies

EXPOSURE   DISEASE   STUDY STARTS

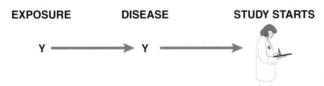

Prevalence surveys

EXPOSURE   DISEASE   STUDY STARTS

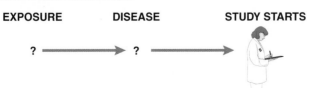

● **Figure 5-3** Summary of chronological sequence in nonexperimental study designs.

ECOLOGICAL STUDIES

An ecological study is any kind of study in which data is collected about a whole population (a large group or community) and is analyzed at that level. Examples would include studies of the rate of diabetes in countries with different levels of automobile ownership, studies of homicide rates in states with different firearms laws, and studies of lung cancer rates in places with different tobacco taxes. They can usually be done quickly and inexpensively using existing bodies of data.

Unlike cohort studies, no data is available or analyzed about individuals; for example, an ecological study may show that communities with large numbers of fast food outlets have higher rates of diabetes, but it would not gather data about whether an individual's food choices relate to his or her risk of diabetes.

Ecological studies need not be nonexperimental: **community intervention trials** are experimental trials done on a population or community-wide basis. The experimental group consists of an entire community, while the control group is an otherwise similar community that is not subject to any kind of intervention. An example is the COMMIT trial, in which the rates of quitting smoking were compared in 11 pairs of North American communities—the "experimental" communities, in which multiple smoking education activities were conducted (such as media campaigns), and the matched control communities, in which no such intervention was performed (Fisher, 1995).

POST-MARKETING STUDIES

A special kind of nonexperimental study is **post-marketing surveillance** (PMS), also known as **Phase 4** or **pharmacovigilance** studies. These involve gathering data about patients taking a particular medication that is already in general use; they may be used to help evaluate the treatment's effectiveness, but in particular they monitor its safety. An adverse effect that occurs in, say, 1 in 5,000 patients, or just in particular kinds of patients, may remain undetected in a clinical trial, but it will become apparent if hundreds of thousands or millions of people are followed in Phase 4 studies, especially as these people will have multiple comorbidities and take multiple other medications. Such studies may be required by the FDA, or may be done by the drug manufacturer for marketing or other purposes. In the United States, these studies are conducted both actively and passively:

- **Active PMS** is done by the FDA's Center for Drug Evaluation and Research's (CDER) Office of Surveillance and Epidemiology, using reviews of hospital or outpatient medical records, pharmacy databases, and laboratory data, or by establishing **registries** (e.g., of all pregnant women who have been prescribed a certain drug).
- **Passive PMS** is done by the FDA MedWatch program, in which doctors and the general public voluntarily report adverse reactions.

These processes may show long-term negative outcomes, rare adverse effects, or even simple problems such as confusion between drugs with similar names, and PMS has led to several drugs being withdrawn from the market, having warnings attached, being limited to certain kinds of patients, or simply being renamed.

Chapter 6

Answering Clinical Questions I: Searching for and Assessing the Evidence

Chapter 5 reviewed various research methods that generate data. This chapter deals with searching for, evaluating, and using research data to help answer clinical questions.

Take the example of an obese but otherwise healthy 50-year-old white American woman with hyperlipidemia; we might want to know what the best treatment is for her. Before looking for studies that might provide an answer, the first task is to clarify the question, which needs to be broken down into four components, commonly summarized by the term *PICO*:

P: Who are the *Participants* (*Patients, Population*)?

It is unlikely that a study has been done on patients who are exactly like the one we are interested in. If we limit our question to such patients, we will reduce or even eliminate the chance of finding relevant evidence; if we broaden it, we risk finding evidence that is invalid for our patient. In this case, we might settle for studies on otherwise healthy female patients with hyperlipidemia, leaving aside for the time being factors such as age, race, obesity, or nationality.

I: What is the *Intervention*?

The intervention in this case may be a drug, a diet, or an exercise or educational program. There is unlikely to be a single study of all of these, so we might choose a narrower question, such as "Is drug treatment for hyperlipidemia beneficial?"

C: What is the *Comparison*?

The comparison may be no treatment (or a placebo), or it may be another drug, a diet, or an exercise program.

O: What is the *Outcome*?

Asking "Is drug treatment beneficial?" is clearly too vague. It is crucial to know whether treatments produce outcomes of clear clinical importance (such as how a patient feels, functions, or survives). Studies that provide this are called **POEMs** ("Patient-Oriented Evidence that Matters"); for example, POEMs for hyperlipidemia treatments would examine outcomes such as death, or disability due to heart attack or stroke. The cost of the treatment (or of the disease) to the patient is also a patient-oriented outcome.

In contrast, studies that measure only physiologic outcomes, or **biomarkers** (such as cholesterol levels, intimal medial thickness, or blood pressure), are termed **DOEs** ("Disease-Oriented Evidence"). These physiologic outcomes are **surrogate outcomes** (or, more properly, surrogate **endpoints**, as they are substitutes for actual outcomes). As it is much easier, cheaper, and quicker to obtain DOEs than POEMs, reliance on surrogate endpoints is very common, particularly for treatments for chronic diseases; indeed, many treatments are approved on the basis of surrogate endpoint data only.

It is often assumed that if a treatment improves a surrogate endpoint, it is beneficial, but this is not necessarily the case. Many treatments that improve surrogate endpoints lack evidence that they improve actual outcomes, and several have been found to actually worsen clinical outcomes, either before or after coming into widespread clinical use. For example, despite producing dramatic improvements in cholesterol profiles in Phase 2 trials, the drug torcetrapib was abandoned in 2007 after an estimated $800M had been spent on its development, following a 15,000-patient Phase 3 trial which showed increased rates of death, heart failure, angina, and revascularization procedures in patients receiving the drug (Barter et al, 2007).

Following the PICO process, the refined question might now be "Do drugs prevent or delay death or disability in otherwise healthy women with hyperlipidemia?"

At this point, we might simply ask a senior physician or a respected expert in the field; however, the principle of **evidence-based medicine** (**EBM**) is that *actual evidence, rather than opinion, must be the primary scientific basis for making clinical decisions*; different types of evidence are judged to be of different value.

Hierarchy of Evidence

The **hierarchy of evidence** (Table 6-1) shows which kinds of information are the most valuable. The best evidence is provided by **evidence syntheses**, with **systematic reviews** at the highest level, followed by **meta-analyses**; next come experimental data from randomized controlled trials (RCTs) and then non-RCTs, followed in turn by nonexperimental or observational data from cohort studies, case–control studies, case series studies, and case reports.

This hierarchy is not absolute; for example, a meta-analysis of well-conducted randomized trials may provide better evidence than a systematic review of poorly conducted nonrandomized trials, and within each level, there may be studies of high or poor quality. Note that the very lowest level comprises expert opinion based only on anecdote, physiologic or *in vitro* data, or first principles, none of which strictly constitute evidence *per se*.

TABLE 6-1	HIERARCHY OF EVIDENCE
Evidence syntheses	Systematic reviews Meta-analyses
Experimental data	RCTs Non-RCTs
Nonexperimental (observational) data	Cohort studies Case–control studies Case series studies Case reports Expert opinion based on anecdote, physiologic or *in vitro* data, or first principles

Systematic Reviews

Unlike the traditional literature review, which typically reviews only the published studies that the authors judge to be most important, **systematic reviews** are objective, with the goal of enabling clinical decisions to be made on the basis of all the good-quality studies that have been done.

In a *bona fide* systemic review, there is a defined clinical question (as above), and the studies reviewed meet specific **inclusion** and **exclusion criteria**, typically using the term *PICOS*, which comprises the PICO parameters with the addition of "S": the type of *Study* design (e.g., RCTs). The literature is searched in a painstaking way to include *all* relevant studies, typically including

- foreign language studies (thus avoiding **language bias**);
- the so-called **gray literature**, including unpublished studies. Published studies are not representative of all the studies on a particular topic for several reasons, including the fact that

negative studies, which fail to find statistical significance, are less likely to be published than **positive studies**, resulting in **publication bias**. Negative or unpublished studies are also likely to take longer to be reported, resulting in **time lag bias**, and they are likely to be less widely reported, cited, reviewed, or publicized, resulting in **reporting**, **citation**, and **dissemination biases**. The "gray literature" may include studies in trial registries (such as ClinicalTrials.gov, with over 100,000 trials; registration is mandatory for trials in the United States) or in the files of pharmaceutical companies, known researchers in the area, and study sponsors, and also theses and studies reported only as conference presentations. Note that standard databases such as MEDLINE, although important, include only published studies;

- a manual "hand search" of journals likely to have published relevant studies, for studies that may have been miscoded or mislabeled and therefore do not appear correctly in electronic search results. This may also reveal data that has been published more than once, which causes **multiple publication bias**;
- studies identified in the citation lists of studies otherwise identified.

Relevant studies are identified, verified as meeting the inclusion and exclusion criteria, and their validity and handling of missing data are then evaluated, ideally by two or more independent reviewers. A final list of acceptable studies is thus obtained, and their data is then abstracted to fit a uniform template, enabling a quantitative analysis to be performed.

META-ANALYSIS

Meta-analysis is the quantitative analysis of the results of a systematic review; its validity depends on the quality of the review and the underlying studies. Advantages of analyzing the combined results of multiple studies include the following:

- With an increased total sample size, the power of the analysis is increased, so Type II errors (in which trials fail to show effects that are actually present) are less likely, and the precision of the estimate of any effects is increased.
- If multiple studies show significant results, Type I errors are less likely (using a p of .05, the probability of one study making a Type I error is .05; the probability of multiple studies all making errors at this level is dramatically lower).
- Multiple studies are likely to have been done in different geographical areas, with patients drawn from different populations, therefore increasing the generalizability of the results.

The extent to which a systematic review is influenced by publication and other biases is often assessed by means of a **funnel plot**. A funnel plot displays the **effect size** (see below) plotted against the sample size.

Normally, there will be more variability in the results of smaller studies than larger studies, and so this plot should look like an inverted funnel, with the larger study results clustering around the overall mean effect more closely than do the results of the smaller studies. If the small studies are simply producing less precise results than the larger ones, as expected, their results should be symmetrically spread around the mean, as shown in Figure 6-1.

If this is not the case, as shown in Figure 6-2, where the distribution of results of smaller studies is asymmetrical, it suggests that some smaller studies that showed unfavorable results may be missing from the analysis because of various biases.

A single estimate of the results of all the studies in the systematic review is obtained by pooling their results; sometimes the individual study results are **weighted**, altering the influence they have on the overall result, according to the precision of their estimate. The result is a measure of **effect size**, typically the **risk ratio** (also known as **relative risk**, discussed further in Chapter 8), which is *the ratio of the frequencies of events in the intervention group versus the control group* (e.g., the frequency of heart attacks in a group of patients given a cholesterol-lowering drug vs. the frequency of heart attacks in a group of patients given a placebo); so a ratio of 1 would reflect no

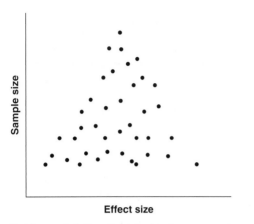

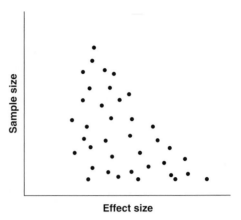

● **Figure 6-1** Funnel plot showing apparent absence of publication bias.

● **Figure 6-2** Funnel plot showing apparent presence of publication bias.

difference, a ratio of 2 would reflect a doubling of the risk with the drug, and a ratio of 0.5 would reflect a halving of the risk.

However, this measure may be misleading if the studies in the systematic review are too different from each other (such as in patient types, the intervention studied, or the outcome measures) to be legitimately combinable. Meta-analyses use **tests of heterogeneity** to test the null hypothesis that the underlying effects seen in the studies are similar and that any difference in the effects is due to chance; if the null hypothesis is rejected, then it is concluded that there are significant differences between the studies that make a single pooled outcome unlikely to be valid.

If this seems to be the case, a **sensitivity analysis** may be done, in which various categories of studies (e.g., unpublished studies, apparently low-quality studies, commercially-sponsored studies) are excluded to see whether the overall conclusions are **robust** or whether they are sensitive to the inclusion of these groups of studies. For example, if the exclusion of studies sponsored by drug companies changes the conclusion of a meta-analysis, the validity of these studies would be open to question.

Results of meta-analyses are commonly presented as **forest plots**, which allow the results to be seen at a glance, with ready identification of outliers. Figure 6-3 is an example of a forest plot produced by a meta-analysis of studies of the effects of statin drugs on the risk of developing ventricular arrhythmias in patients with cardiomyopathy; as shown, each individual study's result (its **point estimate**) is plotted separately:

- The size of the rectangle representing the study is proportional to its sample size.
- The horizontal line on either side of the result shows the 95% confidence interval surrounding the point estimate.
- The location of the result shows whether it favors the intervention (the statin drug) or not; the vertical solid line is the "line of no effect," where there is no difference between the intervention and the placebo (i.e., a risk ratio of 1).
- The overall pooled result of the meta-analysis is shown by a diamond: its center shows the overall aggregate effect, and its lateral corners show the associated 95% confidence interval. If these corners do not cross unity (the line of no effect), then there is an overall statistically significant effect, as the ratio of the frequency of events in the intervention group versus the control group is not 1, at a 95% (or higher) level of confidence, as shown in the figure.

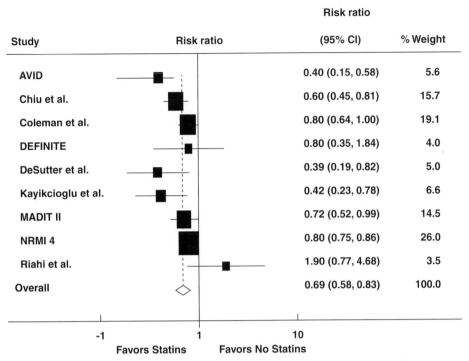

● **Figure 6-3** Forest plot showing the effects of statin drugs on the development of ventricular arrhythmias in patients with cardiomyopathy. (From Wanahita N, Chen J, Bangalore S, et al. The effect of statin therapy on ventricular tachyarrhythmias: a meta-analysis. *Am J Therap.* 2012;19:16–23, with permission.)

SEARCHING FOR EVIDENCE

In searching for evidence to answer a clinical question, the easiest way to find high-quality evidence is by searching for systematic reviews and meta-analyses, particularly using resources devoted to EBM, such as

- Cochrane Database of Systematic Reviews at www.cochrane.org
- TRIP Database at www.tripdatabase.com
- DARE (Database of Abstracts of Reviews of Effects) at www.york.ac.uk/inst/crd/
- Bandolier at www.medicine.ox.ac.uk/bandolier/knowledge.html
- American College of Physicians Journal Club at acpjc.acponline.org/

However, for many clinical questions, evidence syntheses may not be available, and the clinician will need to review the individual studies in the primary literature.

SEARCHING FOR INDIVIDUAL STUDIES

Several tools are available for searching the primary literature, the best known being the PubMed search interface for the MEDLINE database, which currently holds over 21 million publications. No matter what search tool is used, similar principles apply.

As with systematic reviews, searching the primary literature also uses the PICOS concept; each element of PICOS can be framed narrowly or broadly, and the "S" component (study design) of the search will depend on the kind of question being asked.

Clinical questions typically fall into one of five categories—questions about treatment, harm, differential diagnosis, diagnostic tests, and prognosis:

- Questions about treatments ("What is the best treatment for hyperlipidemia?") are best answered by RCTs.
- Questions about harm ("Does a high fat diet increase the risk of cardiac events?") would ideally be answered by RCTs, but for ethical and practical reasons, are often addressed by observational studies such as cohort and case–control studies.
- Questions about differential diagnosis ("What is the cause of this patient's chest pain?") will require studies in which patients with similar signs and/or symptoms undergo different tests (which may include physical examination and a detailed history) and are then followed until the true etiology of their syndrome becomes apparent.
- Questions about diagnostic tests ("What is the best test for the diagnosis of chest pain?") require studies in which the test in question is compared with the results of a "gold standard" criterion for the diagnosis in question (this will be reviewed in Chapter 7).
- Questions about prognosis ("How long is it likely to be before this patient with hyperlipidemia has a cardiac event?") involve following patients with a particular diagnosis, or undergoing a particular treatment, for an extended time, with a form of cohort study. (Methods for assessing prognosis include survival analysis, discussed in Chapter 4.)

Once the question has been clarified, search terms are refined by

- using tools such as the Medical Subject Headings (MeSH) vocabulary; this is a hierarchical thesaurus of over 25,000 defined medical terms. Each article in PubMed has been assigned descriptors from the MeSH vocabulary, and by using appropriate MeSH search terms, more accurate results are obtained;
- broadening the search by including synonyms (e.g., hyperlipidemia and hypercholesterolemia), abbreviations (e.g., RCTs), and different spellings (e.g., anemia, anaemia). The MeSH thesaurus helps with this, as entering a search term in MeSH will show its synonyms, as well as what higher-level terms it is a subset of and what subterms lie under it when it is "exploded";
- limiting the search (such as to RCTs on humans only, to certain publication dates, types of article, or specific ages), either by tools that are built into the search engine or by using Boolean operators (such as AND, OR, and NOT).

By doing this, typically in a series of iterations, search results can be drastically narrowed down or expanded to produce a usable number of citations that are appropriate for the question. In PubMed, the Clinical Queries tool also helps accomplish this in a simplified but faster way (but note that the preset filter it provides for "systematic reviews" uses this term in a very broad way that is not consistent with its usual definition).

APPRAISING STUDIES

Once a search result has been obtained, whether of evidence syntheses or primary studies, it must be appraised. Potential flaws in systemic reviews and meta-analyses have been discussed above. Potential flaws in primary studies (biases, confounders, lack of representativeness, loss to follow-up, problems of internal validity, external validity, and generalizability) have been dealt with in Chapter 5, and those involved in hypothesis testing (Type I and II errors, lack of power, lack of clinical as opposed to statistical significance, inappropriate subgroup analyses) were reviewed in Chapter 3.

APPLICATION TO PATIENT CARE

No matter what kind of research has been used to answer a clinical question, the results of even the best-conducted and most valid study will not be of practical use in patient care unless the treatment or intervention is actually available, and

- it suggests real-world **effectiveness**, and **cost-effectiveness**, on patient-oriented outcomes;
- it shows a **causal relationship** between the independent variable (the treatment, intervention, test, etc.) and the dependent variable.

EFFECTIVENESS VERSUS EFFICACY

Clinical trials are likely to be performed in near-ideal conditions, with careful attention to patient compliance and adherence to treatment protocols, better availability of support staff, little or no (or even negative) cost to the patient, and less time pressure than in regular clinical practice. The ability of a treatment to produce results under these ideal conditions is its **efficacy** ("*can* it work?"), which is likely to be higher than its **effectiveness**—its ability to produce results in real clinical use, where time and resources are likely to be more limited ("*does* it work in the real world?"). For example, the *effectiveness* of a vaccination program for a large rural population in a develop-ing country is likely to be rather lower than its *efficacy* in clinical trials performed in an urban academic medical center.

For the clinician, the difference between efficacy and effectiveness amounts to asking whether the benefits seen in a clinical trial are likely to be reproducible for the patient you have in front of you. The "PICO" parameters again help:

- Are the *Patients* in the clinical trial similar to your patient? Elderly patients, for example, are typically underrepresented in clinical trials, as are patients with multiple comorbidities or on multiple other medications.
- Is the *Intervention* the same? Perhaps the study drug used a different dose, or a form that is similar but not identical to the one available in regular practice for various reasons (unavailability, formulary restrictions, cost, etc.).
- What is the *Comparator*? In clinical practice, there is typically a choice of several treatments, but there may be few or no trials of one treatment against another (so-called "head-to-head" or "active comparator" trials), rather than trials of treatments against a placebo.
- What is the *Outcome*? POEM data is far preferable to DOE data, but may not exist.

COST-EFFECTIVENESS

If an intervention is effective, a critical question is whether it is **cost-effective** ("Is it worth it?"), as medical resources are not unlimited. A new treatment that is cheaper *and* more effective than an existing one is clearly cost-effective, but if a new treatment is more expensive and more effective, or cheaper and less effective, questions of cost-effectiveness arise and will inevitably involve value judgments (see Chapter 7).

A treatment that is judged cost-effective does not necessarily save money, and a treatment that is cheaper is not necessarily cost-effective. Cost-effectiveness is very sensitive to the cost of a treat-ment, which will vary considerably in different places (e.g., a coronary bypass is likely to be much cheaper in India than in the United States) and at different times (a drug may become dramatically cheaper once it goes off-patent and generic formulations are available). A treatment that is highly efficacious in trials will be ineffective in practice if patients, or health care systems, are unable to afford it. (More formal aspects of cost-effectiveness will be discussed in Chapter 7, and the related concept of "number needed to treat" will be discussed in Chapter 8.)

CAUSALITY

Although randomized controlled clinical trials are the gold standard for demonstrating causality, a single RCT is unlikely to be taken as conclusive, especially when applying the PICO questions above. Clinicians are frequently required to use their best judgment about questions for which there are no good evidence syntheses, or only inadequate or inconsistent RCT data, only DOEs but no POEMs, or only correlational or observational data. This is especially true for questions about diet or lifestyle factors or the use of complementary treatments (e.g., supplements, herbal medicines, acupuncture, hypnosis, chiropractic). Evaluating a possible causal relationship between independent and dependent variables involves considering

- the **strength** of the association (such as between use of a treatment and reduction in symptoms);
- the degree to which the findings have been **replicated** consistently in different settings;
- the **specificity** of the association (such as between exposure to a risk and the resulting symptoms);
- **temporal relationships**: is it clear that the proposed cause (risk factor, treatment) came before the outcome (disease, improvement)? If temporality is unclear, it is possible that apparent causation may in fact be **reverse causation**, in which the dependent variable is actually the independent variable, and vice versa. For example, studies in many developing countries have found that breast-fed infants are at greater risk for growth retardation, which suggests that breast-feeding may cause poor growth; but there is evidence that the reverse is actually the case, and that mothers are in fact more likely to breast-feed a child who they perceive as growing poorly;
- **biological gradients**: is there a **dose-response relationship**, such that as exposure increases, the outcome also changes? (E.g., is there a relationship between the lifetime number of cigarettes smoked and the risk of lung cancer, or between the dose of a cholesterol medication and the risk of myocardial infarction, or between the number of chiropractic adjustments and back pain?)
- **plausibility**: does the relationship fit in with our commonsense knowledge of how the world works?
- **coherence**: do the findings fit in with other data or theoretical models to make a coherent picture?
- **analogies**: we are more ready to accept causal explanations that are similar to other explanations that we already accept, for example, "when you take a drug to reduce the cholesterol in your arteries, the effect is like unblocking pipes."

REFERRING PATIENTS TO APPROPRIATE INFORMATION RESOURCES

Answers to clinical questions are of course sought by patients as well as doctors, and the majority of adults (in the United States at least) have sought health information online. Medical information online is almost limitless, and much of it is inaccurate if not misleading, whether for commercial or other reasons. Patients can be advised that the following are generally recognized to be reliable sources of information:

- Medline Plus (medlineplus.gov), the National Library of Medicine's Web site for consumer health information
- Centers for Disease Control and Prevention (www.cdc.gov), particularly for current information on travelers' health worldwide
- Familydoctor.org (familydoctor.org), produced by the American Academy of Family Physicians (AAFP)
- Cancer.gov (www.cancer.gov), the official Web site of the National Cancer Institute
- Healthfinder (www.healthfinder.gov), providing health information from the U.S. Department of Health and Human Services

Patients performing their own searches for medical information online should be advised to

- know the sponsor of the Web site, whether a government agency, professional organization, nonprofit group, commercial organization, or individual patient; each of these may have their own agenda;
- look at the source of the information: are the authors listed, and what are their credentials? Are references to legitimate publications cited?

- be skeptical, especially of claims of "breakthroughs," "cures," "secret ingredients," or "what they don't want you to know." Sites selling "alternative" treatments, self-published books, or unconventional tests, and sites where editorial content and advertising are not clearly distinguished, should be treated with suspicion;
- understand that case histories, anecdotes, testimonials, and unsupported "expert opinion" are all considered the lowest quality of evidence;
- realize that medical information is constantly changing and that a source that does not list the date on which it was produced is likely to be out of date;
- look for a current certification by the HON (Health On the Net) Foundation, a nonprofit organization that certifies that Web sites comply with a number of ethical principles; certified Web sites show an icon that links back to the HON Web site, showing details of the certification.

Chapter 7

Answering Clinical Questions II: Statistics in Medical Decision Making

Medical decision making often involves using various kinds of test data. Any physician using a diagnostic test—whether a physical test or a laboratory test performed on an individual patient, or a screening test being used on a whole population—will want to know how good the test is.

This is a complex question, as the qualities and characteristics of tests can be evaluated in several important ways. To assess the quality of a diagnostic test, as a minimum we need to know its

- validity and reliability
- sensitivity and specificity
- positive and negative predictive values

When using quantitative test results—such as measurements of fasting glucose, serum cholesterol, or hematocrit levels—the physician will also need to know the **accuracy** and **precision** of the measurement as well as the **normal reference** values for the variable in question.

Validity

The **validity** of a test is *the extent to which it actually tests what it claims to test*—in other words, how closely its results correspond to the real state of affairs. The validity of a diagnostic or screening test is, therefore, its ability to show which individuals have the disease in question and which do not. To be truly valid, a test should be highly sensitive, specific (discussed below), and unbiased.

Quantitatively, the validity of a diagnostic or screening test is the proportion of all test results that are correct, as determined by comparison with a standard, known as the **gold standard**, which is generally accepted as correct.

Validity is synonymous with accuracy. As stated in Chapter 2, the accuracy of a figure or measurement is the degree to which it is immune from systematic error or bias, so to the extent that a measurement or test result is free from systematic error or bias, it is accurate and valid. When assessing the validity of a research study as a whole, two kinds of validity are involved:

- **Internal validity**: are the results of the study valid for the sample of patients who were actually studied?
- **External validity**: are the results of the study valid for the population from which the sample in the study was drawn? If the external validity of a study is such that it applies to *other* populations as well, it is said to be **generalizable**.

For example, a randomized controlled trial of vitamin D supplements might demonstrate that they are effective in preventing osteoporosis in a sample of 500 postmenopausal American white women. The study, if properly performed, so that there are no confounding factors, would be internally valid for this sample. It would be *externally* valid only if the sample were representative of all American white postmenopausal women—but the extent to which it is *generalizable* to non-American, non-white, or nonmenopausal women is open to debate.

Reliability

Reliability is synonymous with repeatability and reproducibility: it is *the level of agreement between repeated measurements of the same variable.* Hence, it is also called **test–retest reliability**. When testing a stable variable (such as a patient's weight), it can be quantified in terms of the correlation between measurements made at different times—this is the test's "reliability coefficient."

Reliability corresponds to precision, defined in Chapter 2 as the degree to which a figure is immune from random variation. A test that is affected very little by random variation will obviously produce very similar results when it is used to measure a stable variable at different times. A reliable test is therefore a consistent, stable, and dependable one.

The reliability or repeatability of a test influences the extent to which a single measurement may be taken as a definitive guide for making a diagnosis. In the case of a highly reliable test (such as a patient's weight), one measurement alone may be sufficient to allow a physician to make a diagnosis (such as obesity); but if the test is unreliable in any way this may not be possible. The inherent instability of many biomedical variables (such as blood pressure or blood glucose) often makes it necessary to repeat a measurement at different times, and to use the mean of these results to obtain a reliable measurement and make a confident diagnosis.

In practice, neither validity nor reliability are usually in question in routine medical laboratory testing. Standard laboratory tests have been carefully validated, and quality control procedures in the laboratory ensure reliability.

However, a reliable and precise test or measurement is not necessarily valid or accurate. For example, it would be possible to measure the circumference of a person's skull with great reliability and precision, but this would not be a valid assessment of intelligence!

A reliable and precise measurement may also be rendered invalid by bias. A laboratory balance, for example, may weigh very precisely, with very little variation between repeated weighings of the same object, but if it has not been zeroed properly, all its measurements may be too high, causing all its results to be biased, and hence inaccurate and invalid.

Conversely, a measurement may be valid, yet unreliable. In medicine this is often due to the inherent instability of the variable being measured. Repeated measurements of a patient's blood pressure may vary considerably; yet if all these measurements cluster around one figure, the findings as a whole may accurately represent the true state of affairs (e.g., that a patient is hypertensive).

Reference Values

Even a high-quality, valid, and reliable measurement does not in itself permit a diagnosis to be made—doing this requires knowing the measurement's range of values among normal, healthy people. This range is called the **normal range** or **reference range**, and the limits of this range are the **reference values** (also called the **upper** and **lower limits of normal**, or **ULN** and **LLN**) that the physician will use to interpret the values obtained from the patient. (The range between the reference values is sometimes called the **reference interval**).

How is a valid set of reference values established? The normal range of a biomedical variable is often arbitrarily defined as the middle 95% of the normal distribution—in other words, the population mean plus or minus two standard deviations (explained in Chapter 2). The limits of this range, derived from a healthy population, are therefore the "reference values." This assumes that:

- the 95% of the population lying within this range are "normal," while the 5% beyond it are "abnormal" or "pathologic," and
- the normal range for a particular biomedical variable (e.g., serum cholesterol) can be obtained by measuring it in a large representative sample of normal, healthy individuals, thus obtaining a normal distribution; the central 95% of this normal distribution would then be the "normal range."

Although manufacturers of commercial tests may attempt to establish a reference range by testing thousands, or even tens of thousands of individuals,

- there is nothing inherently pathologic about the 5% of individuals outside this "normal range"; typically, there are some healthy people who have "abnormally" high or low values. Indeed, in some cases an abnormal value—such as a low serum cholesterol value or a high IQ—may be a positive sign rather than a negative one;
- many biologic variables turn out to be skewed rather than normally distributed in the population;
- the sample that is tested to establish the normal range is not usually unambiguously free of disease: it is difficult to find a large group of "normal" people who are healthy in every way;
- if this strictly statistical definition of normality and abnormality were adhered to, all diseases would have the same prevalence rate of 5%.

In practice, the normal range and the corresponding reference values in a laboratory's manual often represent a compromise between the statistically-derived values and clinical judgment, and may be altered from time to time as the laboratory gains experience with a given test. The values must always be interpreted in the light of other factors that may influence the data obtained about a given patient, such as the patient's age, weight, gender, diet, and even the time of day when the specimen was drawn or the measurement made.

Sensitivity and Specificity

Sensitivity and specificity are both measures of a test's validity—its ability to correctly detect people with or without the disease in question. They are best understood by referring to Table 7-1, which shows the four logical possibilities in diagnostic testing:

TP: A positive test result is obtained in the case of a person who has the disease; this is a "true positive" finding.

FP: A positive test result is obtained in the case of a person who does *not* have the disease; this finding is therefore a "false positive" one, which is a type I error.

FN: A negative test result is obtained in the case of a person who *does* have the disease; this is a "false negative" result, which is a type II error.

TN: A negative test result is obtained in the case of a person who does not have the disease; this is a "true negative" result.

SENSITIVITY

The **sensitivity** of a test is its *ability to detect people who do have the disease*. It is the percentage of people with a disease who are correctly detected or classified:

$$\text{Sensitivity} = \frac{\text{number testing positive who have the disease (TP)}}{\text{total number tested who have the disease (TP + FN)}} \times 100$$

TABLE 7-1

		Disease	
		Present	Absent
Test Result	**Positive**	True positive (TP)	False positive (Type I error) (FP)
	Negative	False negative (Type II error) (FN)	True negative (TN)

Thus, a test that is *always* positive for individuals with a given disease, identifying *every* person with the disease, has a sensitivity of 100%. An insensitive test therefore leads to missed diagnoses (many false negative results), while a sensitive test produces few false negatives.

A sensitive test is obviously required in situations in which a false negative result has serious consequences. Thus, high sensitivity is required of tests used to screen donated blood for HIV, for cytologic screening tests (Pap smears) for cervical cancer, and for mammograms.

Very sensitive tests are therefore used for screening *or for* ruling out *disease; if the result of a highly sensitive test is negative, it allows the disease to be ruled out with confidence. A mnemonic for this is "SNOUT"—reminding us that a SeNsitive test with a Negative result rules OUT the disease.*

SPECIFICITY

The **specificity** of a test is its *ability to detect people who do not have the disease.* It is the percentage of the disease-free people who are correctly classified or detected:

$$\text{Specificity} = \frac{\text{number testing positive who do not have the disease (TN)}}{\text{total number tested who do not have the disease (TN + FP)}} \times 100$$

Thus, a test that is *always* negative for healthy individuals, identifying *every* nondiseased person, has a specificity of 100%. A test that is low in specificity therefore leads to many false positive diagnoses, while a test that is highly specific produces few false positives.

High specificity is required in situations in which the consequences of a false positive diagnosis are serious. Such situations include those in which the diagnosis may lead to the initiation of dangerous, painful, or expensive treatments (e.g., major surgery, chemotherapy); in which a diagnosis may be unduly alarming (e.g., HIV, cancer); in which a diagnosis may cause a person to make irreversible decisions (e.g., Alzheimer disease, cancer); or in which a diagnosis may result in a person being stigmatized (e.g., schizophrenia, HIV, tuberculosis).

Very specific tests are therefore appropriate for confirming *or* ruling in *the existence of a disease. If the result of a highly specific test is positive, the disease is almost certainly present. A mnemonic for this is "SPIN"—reminding us that a SPecific test with a Positive result rules IN the disease.*

In clinical practice, sensitivity and specificity are inversely related: an increase in one causes a decrease in the other. This is because test results for people with and without a given disease typically lie on a continuum, overlapping each other, rather than forming two totally discrete groups. The clinician therefore has to select a "cutoff point" to make a diagnostic decision.

For example, the fasting plasma glucose levels of a population might form two overlapping distributions, one of people with diabetes, and one of people without diabetes, resembling those shown in Figure 7-1.

It is apparent that when a fasting glucose test is used to diagnose diabetes, the choice of cutoff point will determine the test's sensitivity and specificity.

The currently accepted cutoff point for the diagnosis of diabetes is a fasting glucose of 126 mg/100 mL. Using this cutoff in the population in Figure 7-1, this test would be about 80% sensitive (about 80% of those with diabetes are identified), with a moderate number of false negatives (people with diabetes who are incorrectly classified as nondiabetic). There are also a moderate number of false positives, so the test is only roughly 85% specific.

Reducing the cutoff to 100 mg/100 mL, as shown in Figure 7-2, would make the test about 96% sensitive, correctly identifying almost every person with diabetes. But at this point it would have a very low specificity, and the number of false positive results would be unacceptably high—many people (roughly 40% of those who do not have diabetes) would be incorrectly diagnosed

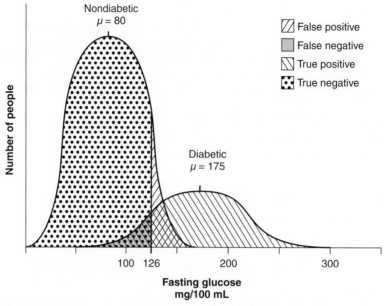

● **Figure 7-1** Diagnosis of diabetes in a hypothetical population, using a cutoff of 126 mg/100 mL.

with the disease. As this suggests, while highly sensitive tests correctly classify the vast majority of people with a certain disease (making few false negative or type II errors), they tend to classify many healthy people incorrectly (making a large number of false positive or type I errors)—so they are likely to have low specificity.

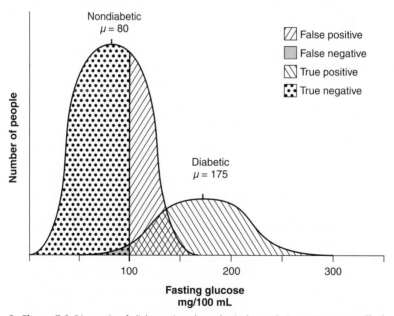

● **Figure 7-2** Diagnosis of diabetes in a hypothetical population, using a cutoff of 100 mg/100 mL.

As the cutoff point is increased, it is clear that the test's sensitivity gradually decreases and its specificity increases, until the cutoff point reaches 160 mg/100 mL, at which point it would be 100% specific, as shown in Figure 7-3, correctly identifying everyone who does not have diabetes.

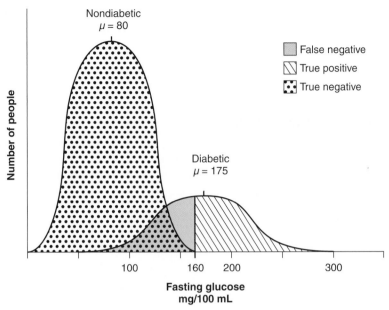

● **Figure 7-3** Diagnosis of diabetes in a hypothetical population, using a cutoff of 160 mg/100 mL.

But it would be highly insensitive, incorrectly classifying around 40% of people with diabetes as being free of the disease. Highly specific tests are therefore likely to be associated with a high number of false negative (type II) errors.

A test can only be 100% sensitive *and* 100% specific if there is no overlap between the group that is normal and the group that has the disease. For example, if nobody with diabetes had a fasting glucose level below 130 mg/100 mL, and nobody without diabetes had a level above 120 mg/100 mL, there would be no problem of a tradeoff between sensitivity and specificity—a cutoff point of 126 mg/100 mL would be perfect. This kind of situation does not occur commonly, and when it does, the disease may be so obvious that no diagnostic testing is required.

Although there are tests of relatively high sensitivity and specificity for some diseases, it is often best to use a combination of tests when testing for a particular disease. A highly sensitive (and usually relatively cheap) test is used first, almost guaranteeing the detection of all cases of the disease, at the expense of including a number of false positive results. This is then followed by a more specific (and usually more expensive) test of the cases with positive results, to eliminate the false positives.

Screening programs in particular will normally require that a highly sensitive test is used initially. Screening involves testing apparently healthy, normal people for the presence of possible disease. If a screening test is negative, the disease is essentially ruled out (according to the SNOUT principle above). Examples include screening donated blood for HIV or hepatitis, Pap tests for cervical cancer, or mammograms for breast cancer, and so on.

The findings of a screening test do not generally diagnose a disease; because screening tests are not usually very specific, a substantial number of false positives are likely to occur. Thus, most women with abnormal mammograms do not have breast cancer, and most people with findings of occult blood in the stool do not have colon cancer; but these findings do mandate a more specific test (such as a breast biopsy or a colonoscopy) to rule in the disease, according to the SPIN principle.

The overall **accuracy** of a clinical test is *the proportion of all tested persons who are correctly identified by the test*, that is, the proportion of all test results, both positive and negative, that are correct.

Accuracy is therefore the number of "true" results (true positives and true negatives) divided by the total of all the test results (true positives, true negatives, false positives, and false negatives); that is,

$$\text{Accuracy} = \frac{(TP + TN)}{(TP + TN + FP + FN)}$$

Receiver Operating Characteristic Curves

To help to determine an appropriate cutoff point for a "positive" test, the relationship between sensitivity and specificity can be clarified by plotting a test's true positive rate (sensitivity) against its false positive rate (100 − specificity) for different cutoff points. Such a plot is a **receiver operating characteristic** (ROC) curve, as shown in Figure 7-4.

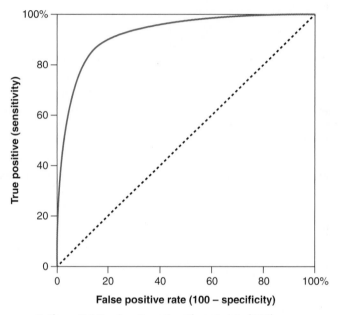

● **Figure 7-4** Receiver Operating Characteristic (ROC) curve.

Each point on the ROC curve represents a sensitivity/specificity pair corresponding to a particular cutoff point or decision threshold. A test that discriminates perfectly between the presence and absence of disease would have an ROC curve that passes through the upper left corner (100% sensitivity, 100% specificity)—so the closer the curve is to the upper left corner, the higher the overall accuracy of the test. A completely random test (e.g., coin tossing) would give an ROC "curve" that is actually the dashed line.

The shape of the curve therefore reflects the quality of the test; the better the test, the more the curve moves to the upper left. This can be quantified in terms of the area under the curve (AUC); the worst case is 0.5 (the dashed line), and the best is 1.00 (upper left-hand corner). A "good" test is one with a high rate of true positives and a low rate of false negatives over a reasonable range of cutoff values; in other words it has a high AUC as the curve moves towards the upper left corner.

Figure 7-5 shows an ROC curve with the three different cutoff points for the diagnosis of diabetes discussed above (point A represents a cutoff of 160 mg/100mL, point B 126 mg/100 mL, and point C 100 mg/100 mL). This curve has an excellent AUC of above 0.9. An alternative test for diabetes, such as a test for urine glucose, would likely have a much lower rate of true positives for a given rate of false positives, as shown in the figure by the dotted line, with a lower AUC. As a rule of thumb, an AUC of 0.5 to 0.6 is almost useless, 0.6 to 0.7 is poor, 0.7 to 0.8 is fair, and 0.8 to 0.9 is very good.

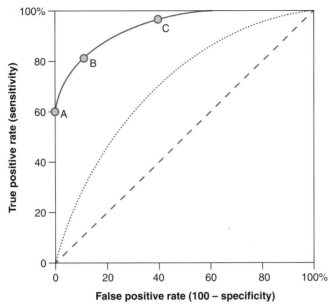

● **Figure 7-5** ROC curves comparing different cutoff points and different tests. An excellent, clinically oriented website, which allows the reader to interactively see the effects of different cutoff points and different situations on type I and type II error rates and ROC curves, is available at http://www.anaesthetist.com/mnm/stats/roc/Findex.htm.

AUC reflects the test's ability to **discriminate** between those with and without disease. ROC curves therefore allow different tests and different cutoff points to be compared.

Predictive Values

When the sensitivity of a test is known, it answers the question, "If the patient has the disease, what is the ability of the test to discover this?" When the specificity of a test is known, it answers the question, "If the patient is free of the disease, what is the ability of the test to discover this?"

These are the kinds of questions that an epidemiologist might ask when screening for a disease. The epidemiologist wants to know, for example, how good a test is at detecting the presence or absence of HIV infection, or what percentage of women with breast cancer will be detected with a mammogram.

However, these are not the kinds of questions that the practicing physician or the patient typically ask; when faced with a test result, they want to know how likely it is that the disease really is present or absent, that is, "If the test is positive (or negative), how likely is it that the patient really has (or does not have) the disease?" This is a different question altogether, and answering it requires knowledge of the **predictive values** of the test.

POSITIVE PREDICTIVE VALUE

The **positive predictive value (PPV)** *of a test is* the proportion of positive results that are true positives, *that is, the likelihood that a person with a positive test result truly has the disease:*

$$PPV = \frac{\text{number who test positive and have the disease (TP)}}{\text{total number who test positive (TP + FP)}}$$

Knowing a test's PPV allows us to answer the question, "Given that the patient's test result is positive, how likely is it that he or she really has the disease?"

NEGATIVE PREDICTIVE VALUE

The **negative predictive value** (**NPV**) *of a test is* the proportion of negative results that are true negatives, *that is, the likelihood that a person with a negative result truly does not have the disease:*

$$NPV = \frac{number\ who\ test\ negative\ and\ do\ not\ have\ the\ disease\ (TN)}{total\ number\ who\ test\ negative\ (TN + FN)}$$

Knowing a test's NPV allows us to answer the question, "Given that the test result is negative, how likely is it that the patient really does not have the disease?" Once again, this is the kind of information a patient is concerned about.

While the sensitivity and specificity of a test depend only on the characteristics of the test itself, predictive values vary according to the prevalence (or underlying probability) of the disease. Thus, predictive values cannot be determined without knowing the prevalence of the disease—they are not qualities of the test per se, *but are a function of the test's characteristics and of the setting in which it is being used.*

The higher the prevalence of a disease in the population, the higher the PPV and the lower the NPV of a test for it. If a disease is rare, even a very specific test may have a low PPV because it produces a large number of false positive results. This is an important consideration because many new tests are first used in high-risk populations, in which a given disease may be quite common. Hence, a test may produce only a few false positive results at first, but when it is used in the general population (in which the disease may be quite rare), it may produce an unacceptably high proportion of false positive results.

> For example, a test for malaria which has very high sensitivity and specificity would be likely to have good positive and negative predictive values in sub-Saharan Africa, but when used in the United States, where the prevalence of malaria approaches zero, it would have an extremely low positive predictive value.

> Another example is given by the CA-125 test for ovarian cancer, which has an unacceptably high false positive rate when used for screening women in general, with a positive predictive value (PPV) of just over 2%, but is very valuable in testing women with a suspicious pelvic mass, with a PPV of about 97%.

It is therefore important to use the right test in the right context. A numerical example makes the relationship between predictive values and prevalence clearer:

Table 7-2 shows the results of a hypothetical diabetes screening program in a community of 1,000 adults, using a test that is 90% sensitive and 99% specific. If this is an inner-city, American community with a large proportion of people at high risk of diabetes, the prevalence

TABLE 7-2

		Disease	
		Present	Absent
Test Result	**Positive**	180 (TP)	8 (FP)
	Negative	20 (FN)	792 (TN)

of diabetes might be 20%, that is, 200 people have diabetes. Because the test is 90% sensitive, 180 of the 200 people with diabetes are detected, leaving 20 false negative results. Because the test is 99% specific, 99% (792) of the 800 nondiabetic people are correctly identified, with 8 false positives.

- The PPV of the test is TP / (TP + FP), or 180 / (180 + 8), which is approximately equal to 0.96: so there is a 96% chance that a person with a positive test actually has diabetes.
- The NPV is TN / (TN + FN), or 792 / (20 + 792), which is approximately equal to 0.98, meaning that a person with a negative test can be 98% sure that he or she does not have diabetes.

However, in a different community of 1,000 adults the prevalence of diabetes might be only 2%, with only 20 people with diabetes. Table 7-3 shows the results of using exactly the same screening

TABLE 7-3

		Disease	
		Present	Absent
Test Result	Positive	18 (TP)	10 (FP)
	Negative	2 (FN)	970 (TN)

test in this community. Because the test is 90% sensitive, 18 of the 20 people with diabetes are detected, leaving 2 false negative results. Because the test is 99% specific, 99% (970) of the 980 nondiabetic people are correctly identified, with 10 false positives.

- The PPV of the test is TP / (TP + FP), or 18 / (18 + 10), which is approximately equal to 0.64: there only a 64% chance that a person with a positive test actually has diabetes.
- The NPV is TN / (TN + FN), or 970 / (2 + 790) which is approximately equal to 0.997: a person with a negative test can be virtually 100% sure that he or she does not have diabetes.

The large difference between the PPVs of this test in the two communities is entirely due to their differing prevalence of diabetes. Note that where the prevalence of the disease is high, PPV is increased, and NPV is decreased.

Because PPV increases with prevalence, one way of improving a test's PPV, and hence avoiding a large number of false positive results, is to restrict its use to high-risk members of the population (as is done with the CA-125 test mentioned above).

Likelihood Ratios

Of course we do not know the actual prevalence of a given disease when we are testing an individual patient in clinical practice. However, we do have some kind of an idea as to how likely it is that the patient has the disease: this is our estimate of the **pretest probability** (also known as **prior probability**) of the disease; it is essentially our estimate of prevalence in a particular situation. We can use this information, together with knowledge of the test's sensitivity and specificity, to help interpret the test's results, and even to decide whether to use the treatment at all in a given situation.

For example, when a woman who has been exposed to strep throat presents with a sore throat, but without fever, tonsillar exudate, or cervical lymphadenopathy, we might guess that the pretest probability of strep throat is about .20. We might also decide that we will only prescribe antibiotics for sore throat if the likelihood of it being strep is 0.50 or more: this is our **treatment threshold**. Would a rapid strep test help us to decide what to do or not?

The concept of **likelihood ratio** (**LR**) helps to answer this question: it is the ratio of the likelihood that a given test result will be found in a patient who has the disease, versus the likelihood that the test result will found in a patient who does *not* have the disease.

A **positive** likelihood ratio (**LR+**) is the ratio of the likelihood of obtaining a *positive* test result in a patient *with* the disease, to this likelihood in a patient who does *not* have the disease (i.e., the ratio of the true positive rate to the false positive rate)—in other words, how much *more* likely do we believe it is that the patient actually has the disease in the light of a *positive* test result?

A **negative** likelihood ratio (**LR−**) is the ratio of the likelihood of obtaining a *negative* test result in a patient *with* the disease, to this likelihood in a patient who does *not* have the disease (i.e., the ratio of the false negative rate to the true negative rate)—in other words, how much *less* likely do we believe it is it that the patient has the disease in the light of a *negative* test result?

Table 7-4 illustrates this in a group of 100 people, of whom 10 have strep throat:

TABLE 7-4

		Strep Throat	
		Present	Absent
Test Result	Positive	9 (TP)	10 (FP)
	Negative	1 (FN)	80 (TN)

- Out of the 10 people with strep throat, 9 have positive results on the rapid strep test (this is the test's sensitivity, 90%).
- Out of the 90 people without strep throat, 80 have negative test results on the test (this is the test's specificity, about 89%), leaving 10, or about 11% of those without the disease, with positive results (note that this 11% represents 100% minus the test's specificity).
- The ratio of the likelihoods of getting a positive result in those with the disease versus in those without the disease is therefore 90/11, or about 8: this is the test's positive likelihood ratio, or LR+. The formula for LR+ is therefore

$$LR+ = \frac{\text{sensitivity}}{1 - \text{specificity}}$$

Similarly, out of the 10 people with strep throat, only 1 patient has a negative test result, so the likelihood of obtaining a negative test result is 10% (note that this represents 100% minus the test's sensitivity). Among the 90 people without strep throat, the likelihood of obtaining a negative result (80 people) is 89% (note that this is the test's specificity). The ratio of the likelihoods of getting a negative result in those with the disease versus in those without the disease is therefore 10/89, or about 0.1. The formula for LR− is therefore

$$LR- = \frac{1 - \text{sensitivity}}{\text{specificity}}$$

After obtaining our test result, we can use these likelihood ratios to estimate how likely it is that our patient with sore throat has strep throat—this is the **post-test probability** of the disease.

Calculating post-test probability is a little cumbersome, so in practice it is easiest to use the widely-reproduced nomogram shown in Figure 7-6, or an online calculator.

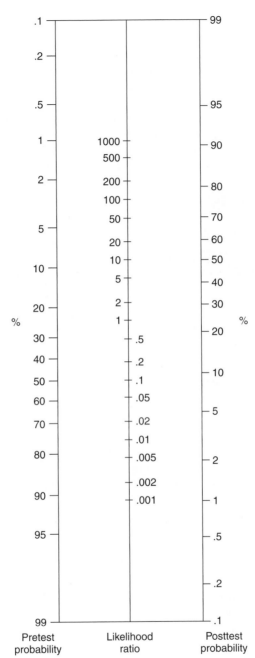

● **Figure 7-6** Nomogram for determining posttest probability (adapted from Fagan, TJ. Nomogram for Bayes's theorem. *New Engl J Med.* 1975;293:257).

The nomogram is used by lining up a straight edge from the pretest probability to the likelihood ratio; the post-test probability is then read in the third column.

In the strep throat example, the pretest probability was .2 (20%), the LR+ was 8, and the nomogram shows a post-test probability of about 70%—in other words, a positive test result for this patient will increase our confidence that she has strep throat (from our initial estimate of 20%) to 70%, which will lead us to prescribe an antibiotic. In this situation, therefore, the result meets the treatment threshold—the level at which it will lead us to initiate a treatment.

Conversely, if she had a negative test result, the nomogram shows that the LR− of 0.1 produces a post-test probability of only 2%.

Likelihood ratios are therefore an indication of the ability of a test to change our assessment of whether a disease is or is not present. Essentially, they combine information about sensitivity and specificity into a single, clinically useful, measure. Because they only use information about sensitivity and specificity, they do *not* depend on the prevalence of the disease, but are fixed, independent characteristics of the test itself (unlike predictive values, as shown above).

Different tests have different likelihood ratios, and the clinical implications of these vary, as shown by the "rules of thumb" for different LRs in Table 7-5. Clearly, a test with a high LR+ is useful to rule in a disease, while a test with a low LR− is useful to rule out a disease.

TABLE 7-5

Ability of test to change posttest probability	LR+	LR−
Excellent	10	0.1
Very good	6	0.2
Fair	2	0.5
Useless	1	1

Table 7-6 summarizes the calculations of sensitivity, specificity, positive and negative predictive values, accuracy, and likelihood ratios.

TABLE 7-6

		Strep Throat	
		Present	Absent
Test Result	Positive	True positive (TP)	False positive (FP)
	Negative	False negative (FN)	True negative (TN)

Sensitivity $= TP / (TP + FN)$;
Specificity $= TN / (TN + FP)$;
PPV $= TP / (TP + FP)$;
NPV $= TN / (TN + FN)$;
Accuracy $= (TP + TN) / (TP + TN + FP + FN)$;
LR+ $=$ *sensitivity* $/ (1 -$ *specificity*$)$;
LR− $= (1 -$ *sensitivity*$) /$ *specificity*.

Prediction Rules

Likelihood ratios are not just a property of individual medical tests; a whole constellation of signs, symptoms, and test results can be used together to help change our pretest assessment of a patient. This is done by **prediction rules**, which quantify multiple clinical findings in order to predict the patient's likely diagnosis, prognosis, or response to treatment.

Prediction rules (sometimes also called **decision rules**) are widely used in daily clinical practice; some of the best-known ones are:

- the CAGE questionnaire (having felt a need to Cut down alcohol intake, having been Annoyed about criticism of drinking, having felt Guilty about drinking, having had an Eye-opener drink in the morning) for the identification of alcohol abuse
- the Ranson criteria for predicting mortality from acute pancreatitis

- the MELD (Model for End-Stage Liver Disease) for predicting a patient's risk of dying while awaiting a liver transplant
- the CHADS2 (Congestive heart failure, Hypertension, Age, Diabetes, Stroke or Systemic emboli) score for predicting the risk of stroke in patients with untreated atrial fibrillation

Prediction rules are developed by identifying a group of patients who are at risk of a specific disease (such as alcoholism) or outcome (such as death from pancreatitis, death from liver disease, stroke from atrial fibrillation), and then recording data from the patient's history, physical exam, and test results which are available in the early stages of the disease. The ability of this data to predict the results of subsequent more detailed testing or follow up of the patient is then evaluated.

After development, prediction rules need to be evaluated:

- When applied in clinical practice, does the rule produce a result that is different from the pretest probability (of alcoholism, death from liver disease, or stroke from atrial fibrillation)? For example, a score of 3 on the CAGE tool has been found to have an LR+ of 13 for diagnosing alcoholism.
- Is the rule generalizable? It should be validated by applying it to different groups of patients in different settings.
- Is the rule useful? It is only useful if it produces results that are better than those produced by physicians' regular clinical judgment.

Clinical judgment will always be required to determine whether a prediction rule should be applied. Hundreds of prediction rules are available in software form, which can be used with smartphones and other devices at the point of care; but the fact that a software application produces a result (for example, that the patient is at low risk of a thromboembolic stroke due to atrial fibrillation, and that anticoagulation is not indicated) does not reduce the physician's clinical, ethical, or medicolegal responsibility for deciding if, when, and how to use the rule and its result.

Decision Analysis

While prediction rules are typically used by physicians at the bedside, broader medical decisions are also made by patients, by the person or organization paying for the treatment (the patient, an insurance company, or a government body, among others), and by professional or regulatory bodies producing clinical guidelines or policies.

A process for incorporating the risks, benefits, and costs of a range of possible treatments in medical decision making is offered by **decision analysis**, which is a formal method for breaking down complex decisions into small components and then analyzing these parts and combining them into an overall framework, which is shown in the form of a **decision tree**.

A decision tree is a diagram which depicts choices, decisions, and outcomes in chronological order from left to right. A square is used to represent a choice to be made (a "decision node"); circles represent possible events ("chance nodes") resulting from the choice by chance; triangles ("terminal nodes") represent outcomes. Probabilities (ranging from 0 to 1, as always) are assigned to each of the possible events; these figures are usually derived from the medical literature.

Figure 7-7 shows a hypothetical decision tree for the choice between medical management and surgery (carotid endarterectomy) for the treatment of asymptomatic carotid artery stenosis. Surgery may result in one of two immediate possible events:

- No intraoperative cerebrovascular accident (CVA; $p = .93$)
- Intraoperative CVA ($p = .07$; note that these two probabilities add up to 1.00)

The patient who undergoes surgery and does not have a CVA still has some risk of a future CVA ($p = .03$, as shown).

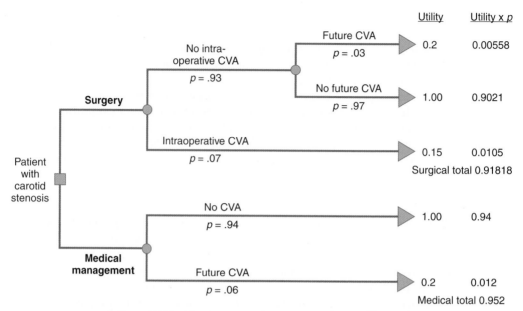

● **Figure 7-7** Decision tree comparing treatments for carotid stenosis.

Medical management may also produce one of two outcomes:

- continuing unchanged health ($p = .94$)
- an eventual CVA ($p = .06$).

Next, a measure of relative value or **utility** is assigned to each possible outcome. Death is normally assigned a utility of 0, complete health a utility of 1, and between those two extremes a utility can be assigned taking into account any kind of outcome, such years of life gained, cost, or the quality of the person's life. The value which is assigned will vary according to the decisionmaker; an insurance company might place a higher value on direct cost savings; a government agency on overall public health outcomes; one patient might place great value on the length of his or her life, while another might place more value on the quality of that life.

As an example, Figure 7-7 assumes that an intraoperative CVA is viewed as a very poor outcome, with a utility of 0.15; while a later CVA is viewed only a little less poorly with a utility of 0.2, and continuing good stroke-free health is given a utility of 1. Likewise, a later CVA with medical management is also given a utility of 0.2.

The total expected utility of each choice is then found by adding up the utilities of each of the outcomes that result from that choice, weighted by their respective cumulative probabilities. For example, for the choice to undergo surgery,

- the utility of 0.15 for an intraoperative CVA, weighted by its probability of .07, is $0.15 \times 0.07 = 0.0105$.
- the utility of 1 for the outcome of no intraoperative CVA and no future CVA, weighted by the probability of .93 for no intraoperative CVA and the probability of .97 for no future CVA, is $1 \times .93 \times .97 = 0.9021$.
- the utility of 0.2 for no intraoperative CVA followed by a future CVA, weighted by their respective probabilities of .93 and .03 is $0.2 \times .93 \times .03 = 0.00558$.
- the sum of these utilities ($0.0105 + 0.9021 + 0.00558$) is 0.91818: this is the total expected utility of surgery.

Similarly, the total expected utility of the choice to have medical management is (0.94×1) for no CVA, plus (0.06×0.2) for a future CVA, $= 0.952$, which is higher than the total for surgery; so with the relative values this decisionmaker has given to each possible outcome, medical management is the best choice.

A different scenario, in which the decisionmaker places a different value on a future CVA, or in which the surgeon has a different rate of intraoperative CVAs, might produce a different best choice. Clearly there are many other factors that could be incorporated in this analysis, such as the likely severity of the CVAs, the severity of the patient's carotid stenosis and his or her prior history of strokes, the risk of other surgical complications, the risk of side effects of medications, the individual patient's operative risk, the risk of multiple events, the timing of events, cost, etc.

Even if both treatment choices may result in a stroke, the one which delays a stroke for several years, or which results in a less disabling stroke, is preferable.

One way of including considerations of duration and quality of life (and therefore including questions of the timing of the outcomes which a treatment might produce) is through the concept of **Quality Adjusted Life Years** (**QALYs**)—a measure of disease burden which takes the quality as well as quantity of life into account.

Once again, this involves placing a value on a given state of health, from 1 (full health) to 0 (death), and this value is used to adjust the years of remaining life that the patient might expect, on the assumption that life spent while impaired (whether through pain, disability, or other symptoms) is of less value that time spent without impairments. For example:

- if one choice produced a mean of 10 stroke-free years (with a quality of life of 1) followed by a fatal stroke, it would produce 10 QALYs
- if the other choice produced a mean of 6 stroke-free years, followed by a stroke which produced only minimal deficits (with a quality of life rated at 0.9) for 8 years, followed by a fatal stroke, it would produce $(6 \times 1) + (8 \times 0.9) = 13.2$ QALYs

These QALY figures can be used as the utility figures in a decision analysis.

QALYs can also be used to help determine the economic value of a treatment. Chapter 8 will discuss the concept of number needed to treat (NNT), and how this can be used to determine the cost of (for example) saving one life with a given treatment. Rather than simply looking at the cost of saving a life, we can also look at the cost-per-QALY of a treatment.

> For example, the FREEDOM trial (Cummings et al, 2009) showed that a new drug for osteoporosis, when given annually to osteoporotic women between 60 and 90 years of age, produced a 40% reduction in the relative risk of hip fracture over a 3-year period, but only a 0.5% reduction in absolute risk. Therefore 200 women (the NNT) had to be treated for 3 years to prevent one hip fracture. As the cost of the treatment was about $4950; the cost of preventing one hip fracture was therefore $200 \times 4950 = \$990,000$.

> If we assume that the women in this study had an average remaining life expectancy of 20 years, and that this single 3-year course of treatment prevented this hip fracture from ever occurring, permitting otherwise normal health, the drug produced 20 QALYs; its cost-per-QALY was therefore $49,500 ($990,000 divided by 20).

> Alternatively, if the drug merely delayed a hip fracture for 3 years (producing 3 QALYs), so that the fracture occurred as soon as the treatment was stopped (producing a subsequently impaired and shortened life)—say 5 years with disability, rated with a quality of 0.6, i.e., 3 QALYs), it would have produced only 6 QALYs, at a cost-per-QALY of $165,000.

Judgments of the quality of a person's life will vary subjectively from one person to another, and decisions regarding cost-effectiveness may be ethically and financially difficult; however, unless medical resources are infinite, these decisions have to be made, and decision analysis and QALYs offer an explicit rational framework for this. Few would argue, for example, that a treatment that prevents death and offers the prospect of a normal, healthy life expectancy in children (i.e., it has very low cost-per-QALY) is more valuable than one that merely delays impairment in the very elderly at a high cost-per-QALY.

While it is impractical to use decision analysis for each individual patient's choices, it often plays an important role in the development of evidence-based recommendations, such as **practice guidelines** and **recommendations**. Such guidelines may be invaluable, but their objectivity, quality, and applicability and to a given clinical situation cannot be taken for granted.

Guidelines are developed and published by innumerable bodies, including specialty societies (e.g., the American Academy of Pediatrics), government agencies (such as the U.S. Preventive Services Task Force, USPSTF), international organizations (such as the World Health Organization), academic groups, hospitals, insurance companies and other third-party payers; their production and publication may be funded by many different organizations, some of which (such as pharmaceutical companies and specialty societies) may have a financial interest in certain courses of action. Guidelines may often be inconsistent with each other (e.g., current guidelines regarding prostate cancer screening differ between the American Urological Association, the American Cancer Society, the American College of Surgeons, and the USPSTF).

Assessing a guideline's validity requires some knowledge about the way it was developed:

- Was it developed to answer explicit and important clinical questions and to meet a true need, or are the questions unclear; are there suspicions that the content, or even the mere existence, of the guideline or recommendation has been influenced by other factors—such as a desire to increase or decrease the use of certain tests or treatments?
- Is the guideline based on a systematic review of available evidence, conducted in an explicit and transparent way?
- What is the quality of the evidence supporting the guideline? This may range from very low (unsystematic clinical observations such as case reports, or unsupported expert opinion), to high (consistent data from well-performed randomized clinical trials)
- How good is the evidence linking the treatment options to actual outcomes?

Guidelines and recommendations vary considerably in quality, including in the degree to which they are based on good evidence. The most widely-used method of appraising and rating the quality of practice guidelines is the GRADE (Grading of Recommendations Assessment, Development, and Evaluation) system, which has been adopted by many organizations, including the WHO and the American College of Physicians. The GRADE system rates the strength of recommendations as "strong" or "weak," based on the quality of evidence, the uncertainty about the balance between desirable and undesirable effects, uncertainty or variability in values and preferences, and uncertainty about whether the recommendation would result in a wise use of resources.

Note that high-quality evidence may result in a weak recommendation, if the evidence shows that a treatment's benefits scarcely outweigh its undesirable effects; but even low quality evidence can occasionally produce a strong recommendation (pending the production of better evidence) if a treatment's benefits appear to clearly outweigh its disadvantages.

Assessing the applicability of a guideline or recommendation to a particular situation involves asking the following:

- Does it apply to the kind of patient in the current situation? The guideline's applicability will be limited if the patients in the studies that contributed to its development are dissimilar to the current patient in terms of race, age, gender, disease severity, comorbidities (such as diabetes, obesity, smoking, substance abuse), ability to comply, context (rural, urban, developing country, lack of social support), and so on.

- Does it consider all the treatment options? For example, carotid stents offer an alternative treatment not considered in Figure 7-7.

- Does it consider all the possible outcomes, especially patient-oriented ones (such as quality of life, pain, inconvenience) and not just disease-oriented ones?

- Are the values and utilities clearly specified, and are the patient's values and utilities likely to be similar to those? Some patients may prefer death to severe disability, others the converse.

- If costs are considered, are these costs incurred by patients, insurers, or other stakeholders in the healthcare system? Do the cost estimates apply in the setting in which the guideline is being applied? (Costs of treatments may vary dramatically from one place and time to another.)

- How recent is the guideline? Treatment options, data, costs, or contexts may have changed since the guideline was produced.

Chapter 8

Epidemiology and Population Health

Epidemiology is the study of the distribution, determinants, and dynamics of health and disease in groups of people in relation to their environment and ways of living. This chapter reviews epidemiologic measures of the overall health of groups of people, followed by measures of risk, and the epidemiologic approach to episodes of disease.

Epidemiology and Overall Health

General epidemiological descriptors of the overall health of groups, and especially of entire populations, include *population pyramids*, *mortality*, *life expectancy*, and *infant mortality*, among other measures. These are commonly used to compare the health of populations in different areas or different countries.

POPULATION PYRAMIDS

The age structure of a population has profound effects on present and future public health, and is of great interest to health planners. **Population pyramids** are histograms showing the percentage of a population that falls in different age groups (typically 5-year cohorts)—but unlike a conventional histogram, they are turned on their side, and male and female data are shown on opposite sides of the vertical axis.

They may resemble a pyramid in the case of the population of a developing country with a large number of births, a fairly high infant-mortality rate, and a relatively low life expectancy, as shown for the case of Angola in Figure 8-1.

However, highly developed countries, such as Japan and the United States, with lower birth rates, low infant mortality, and high life expectancy produce a different "pyramid," as shown in Figure 8-2.

Figure 8-2 is a "constrictive" histogram, illustrating an aging population; it portends future problems for health care financing, when an aging population, with increasing needs for medical services, becomes dependent on the productivity of a progressively shrinking proportion of the population who are of working age. The ratio of the productive population (nominally 20 to 64 years old) to the nonworking population (under 20 and above 64 years old) is called the **dependency ratio**:

$$\text{Dependency ratio} = \frac{(\text{population} < 20 \text{ years old}) + (\text{population} > 64 \text{ years old})}{\text{population between 20 and 65 years old}} \times 100$$

Similarly, the **youth dependency ratio** is the ratio of the under-20s to the 20- to 64-year-olds, and the **old age dependency ratio** is the ratio of the over-64s to the 20- to 64-year-olds. The population pyramids illustrate that the old age dependency ratio in Japan is very different from that of Angola.

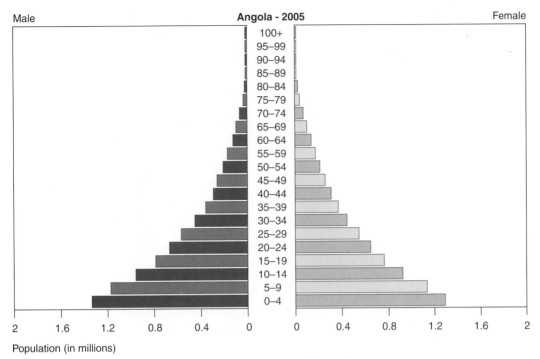

● **Figure 8-1** Population pyramid for Angola, 2005.

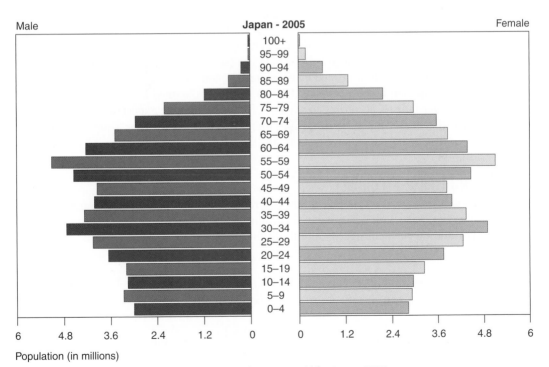

● **Figure 8-2** Population pyramid for Japan, 2005.

Measures of Life Expectancy

Life expectancy is the average number of years of life that a person is expected to live. It is derived from a life table analysis (discussed in Chapter 4). Among cohorts of people who have already died (such as people born before 1890), life expectancy figures are as accurate as the raw data they are based on; for people currently living, they are based on projections, usually with the assumption that current mortality rates remain unchanged. Possible future advances in medicine, or broader socioeconomic changes, produce uncertainty in these projections.

Life expectancy at birth is likely to be quite unrepresentative of the life expectancy of adults, especially in developing countries or in historical cohorts, where a high infant mortality rate results in a very skewed distribution of ages at death. In such situations, life expectancy at (say) 5 years of age may be dramatically higher than life expectancy at birth; as a result, *age-specific* life expectancies are likely to be much more informative. Life expectancies can also be estimated for men versus women, smokers versus nonsmokers, different races, and so on; these estimates are of particular interest for health planners and life insurance companies.

The impact of premature death can be described in terms of **years of potential life lost** (**YPLL**, also called **years of life lost**, or **YLL**). For example, in a population in which a 40-year-old has an average life expectancy of 70, death at the age of 40 would result in 30 years of potential life lost. This figure reflects the effects of changes (whether socioeconomic, in medical care, or public health) according to the ages of the people that they affect; providing defibrillators in nursing homes, for example, is likely to produce a smaller reduction in YPLL than is providing defibrillators at sporting venues, all other things remaining equal.

Other measures introduce consideration of the quality as well as quantity of expected life. A disease such as osteoarthritis, which is not fatal in itself, may produce little or no YPLL; but it may produce significant and prolonged disability, which is described by **disability-adjusted life years** (**DALYs**). DALYs are calculated as the sum of years of potential life lost (YPLL) plus years lived with disability (YLD). As such, one DALY reflects the *loss* of one year of expected healthy life (such as from disease), and DALYs are therefore the converse of quality-adjusted life years (QALYs), discussed in Chapter 7, which reflect a *gain* of years of life (such as from a treatment).

Life expectancy figures can be refined by considering quality of life and the effects of disability, to produce **health adjusted life expectancy (HALE)**, which is obtained by subtracting years of ill health (weighted, like QALYs, according to severity) from standard life expectancy. Like life expectancy, HALE can also be calculated for people of different ages. Osteoarthritis, again, may not affect life expectancy, but may have a major effect on HALE.

Measures of Disease Frequency

Many of the epidemiological terms that follow are **rates**, which all consist of a numerator (usually the number of people with a particular condition) and a denominator (usually the number of people at risk), and they usually specify a unit of time.

MORTALITY

Mortality is the *number of deaths*. The mortality rate is the ratio of the number of people dying (whether of all causes or of a specific disease) to the total number of people at risk:

$$\text{Mortality rate} = \frac{\text{total number of deaths}}{\text{total number of people at risk}} \text{ per unit of time}$$

Mortality rates may be expressed as a percentage, or as the number of deaths per 1,000 or 100,000 people, typically per annum.

Infant mortality is the *number of deaths of infants under one year of age*; the infant mortality rate is the ratio of this to the number of live births, usually expressed per 1,000 live births per annum:

$$\text{Infant mortality rate} = \frac{\text{total number of deaths under 1 year of age}}{\text{total number of live births}} \text{ per annum}$$

Similarly, the under-5 mortality rate is the number of deaths of children between birth and 5 years of age, per 1,000 live births per annum.

The **case fatality rate** (**CFR**) is the ratio of the number of people dying in a particular episode of a disease to the total number of episodes of the disease, expressed as a percentage:

$$\text{CFR} = \frac{\text{number of people dying in an episode of the disease}}{\text{total number of episodes of the disease}} \times 100$$

The CFR is a measure of the prognosis, in terms of life or death, of an episode of a disease, because it shows the likelihood that one episode of it will result in death. It is used to follow the effectiveness of treatments over time or in different places (such as the CFR of acute myocardial infarction or pulmonary embolism this year compared to 20 years ago, or in hospital A vs. hospital B).

ADJUSTMENT OF RATES

Mortality rates are commonly compared across different populations. However, if the populations have different age structures, comparisons will be biased.

> Alaska's mortality rate in 2008 was 507.5 per 100,000 of the population; Florida's was 931.2. Can we conclude that living in Alaska is conducive to longevity? No—Alaska has a much younger population than Florida, so Alaskans would clearly be less likely to die in any given year, other things remaining equal. The mortality rate alone therefore tells us nothing about any real underlying difference in the rate of death between the states.

This biasing influence of a confounding variable, such as age, can be removed by the technique of **adjustment** (or **standardization**) of rates. This involves calculating rates for the two populations as if they were both the same in terms of the confounding variable (in this case, age), often by applying the rates to a standardized population (such as to the age structure of US population as a whole). This kind of process of adjustment can be done not only for age, but also for any other relevant factor that differs between two populations that are being compared. Alaska's 2008 **age-adjusted** (or **age-standardized**) mortality rate was 739.6 per 100,000; Florida's was 679.0—it seems that when age is taken into consideration, Floridians are actually less likely to die in any given year, contradicting the impression given by the **crude** (unadjusted) mortality rate (US Department of Health and Human Services, 2011).

Comparing adjusted mortality rates in a given community with the mortality rate in the population as a whole gives us an idea of whether the rate of death, either overall or from a particular disease, is truly higher in that community. This ratio of actual to expected deaths (adjusted by age and gender) is the **standardized mortality ratio**, or **SMR**. An SMR of 1 means that the actual rate of deaths is the same as the expected rate. For example, if the SMR for motor vehicle accidents in Florida was 1.5, this tells us that the rate of death from these accidents in Florida is 1.5 times the expected rate (in the United States as a whole), even after taking Florida's age and gender structure into account—which would lead us to ask what other factors could account for this difference.

INCIDENCE

The **incidence** of a disease is the *number of new cases occurring* in a particular time period (such as a year). The incidence rate is therefore the ratio of new cases of the disease to the total number of people at risk:

$$\text{Incidence rate} = \frac{\text{number of new cases of the disease}}{\text{total number of people at risk}} \text{ per unit of time}$$

The incidence rate is often stated per 100,000 of the population at risk, or as a percentage. Incidence rates are found by the use of cohort studies, which are therefore sometimes also known as incidence studies (see Chapter 5). For example, if the incidence of shingles in a community is 2,000 per 100,000 per annum, this tells us that in 1 year, 2% of the population experiences an episode of shingles.

Note that *mortality* is actually a special form of incidence, in which the event in question is death rather than contracting a disease. In a disease that is always fatal, mortality and incidence will ultimately be the same. Mortality figures are likely to be more accurate than incidence figures, because deaths are always recorded, while episodes of illness are not.

PREVALENCE

The **prevalence** of a disease is the *number of people affected by it* at a particular moment in time. The prevalence rate is therefore the ratio of the number of people with the disease to the total number of people at risk, which is usually taken to be the total population:

$$\text{Prevalence rate} = \frac{\text{number of new cases of the disease}}{\text{total number of people in the population}} \text{ at a particular time}$$

Like incidence rates, prevalence rates are often stated per 100,000 people, or as a percentage. They are generally found by prevalence surveys. For example, at a given time, 80 out of 100,000 people in a given community might be suffering from shingles, giving a prevalence rate of 0.08%.

Prevalence is an appropriate measure of the burden of a relatively stable chronic condition (such as hypertension or diabetes). However, it is not generally appropriate for acute illnesses, as it depends on the average duration of the disease—it is of little value to speak of the prevalence of pulmonary emboli or myocardial infarctions.

Prevalence is equal to the *incidence multiplied by the average duration of the disease,* so an increased prevalence rate may merely reflect increased duration of an acute illness, rather than suggesting that members of the population are at greater risk of contracting the disease.

The incidence and prevalence rates of shingles given in the above example suggest that the average episode of this illness lasts about 2 weeks, as the prevalence is 1/25th of the annual incidence. If a new treatment cut the duration of an episode of shingles in half, to about 1 week (1/50th of a year), but did nothing to prevent shingles from occurring, the incidence would not change, but the prevalence at any given time would be cut in half:

Before new treatment:

$$\text{Prevalence} = \text{annual incidence} \times \text{average duration (in years)}$$
$$= 2\% \times 1/25$$
$$= 0.08\%$$

After new treatment:

$$\text{Prevalence} = \text{annual incidence} \times \text{average duration (in years)}$$
$$= 2\% \times 1/50$$
$$= 0.04\%$$

Incidence and prevalence are both measures of **morbidity**, or *the rate of illness*. An **epidemic** is the occurrence of a number of cases of a disease that is *in excess of normal expectancy*. Although it is typically applied to infectious diseases (such as an influenza epidemic), it can be applied to any kind of disease (such as an epidemic of obesity). A **pandemic** is an epidemic that *affects a wide geographic area*. The term **endemic** is used to describe a disease that is *habitually present in a given geographic area*, while **hyperendemic** refers to *persistent high levels of a disease*.

THE "EPIDEMIOLOGIST'S BATHTUB"

The relationships between incidence, prevalence, and mortality in any disease can be visualized with the aid of the "epidemiologist's bathtub," shown in Figure 8-3.

- The flow of water through the faucet into the bathtub is analogous to incidence, representing the occurrence of new cases of the disease in the population at risk.
- The level of water in the tub represents the prevalence, or number of cases of the disease existing at any given point in time.
- The flow of water out through the drain represents mortality.
- The evaporation of water represents recovery, whether spontaneous or through treatment, or the emigration of people with the disease (so they are no longer part of the population in question).

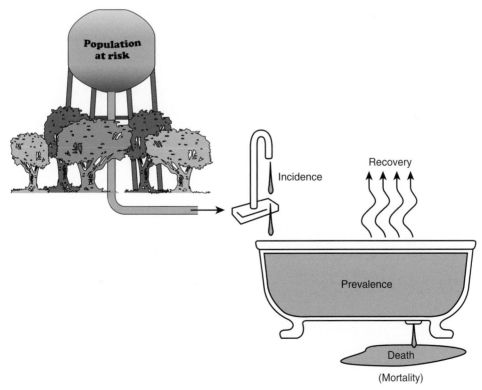

● **Figure 8-3** The epidemiologist's bathtub.

Alzheimer disease provides an example of the application of the "bathtub." If the incidence (inflow of water) of Alzheimer disease is roughly constant, and modern medicine is able to keep Alzheimer patients alive for longer, the rate of mortality from the disease will be reduced (partially blocking the "drain"). However, because there is no cure for the disease, and recovery never occurs, the result is clearly an increased prevalence (increased water level)—which is apparent in the United States today. This picture is also broadly true for HIV and many other chronic diseases in the United States.

Measurement of Risk

Information about the risk of contracting a disease is of great value. The knowledge that something is a risk factor for a disease can be used to help:

- *Prevent the disease.*
- *Predict its future incidence and prevalence.*
- *Diagnose it (diagnostic suspicions are higher if it is known that a patient was exposed to the risk factor).*
- *Establish the cause of a disease of unknown etiology.*

ABSOLUTE RISK

The fundamental measure of risk is incidence. The incidence of a disease is, in fact, the **absolute risk** of contracting it. For example, if the incidence of a disease is 10 per 1,000 people per annum, then the absolute risk of a person actually contracting it is also 10 per 1,000 per annum, or 1% per annum.

It is useful to go beyond absolute risk and to compare the incidence of a disease in different groups of people, to find out if exposure to a suspected risk factor (such as smoking cigarettes) increases the risk of contracting a certain disease (such as lung cancer). A number of different comparisons of risk can be made, including **relative risk**, **attributable risk**, and the **odds ratio**. All these are called **measures of effect**—they measure the effect of being exposed to a risk factor on the risk of contracting a disease.

The ideal way of determining the effect of a risk factor is by a controlled experiment, but as this is rarely ethical, the best alternative is the cohort (prospective) study, in which the incidence of disease in exposed and nonexposed people can be observed directly.

One of the main goals of these studies (such as the Framingham Study described briefly in Chapter 5) is to find the extent to which the risk of contracting the disease is increased by exposure to the risk factor. The two measures that show this are **relative risk** and **attributable risk**.

RELATIVE RISK

Relative risk (RR) *states by how many times exposure to the risk factor increases the risk of contracting the disease. It is therefore the ratio of the incidence of the disease among exposed persons to the incidence of the disease among unexposed persons:*

$$\text{Relative risk} = \frac{\text{incidence of the disease among persons exposed to the risk factor}}{\text{incidence of the disease among persons not exposed to the risk factor}}$$

For example, Table 8-1 reports the results of a hypothetical cohort study of lung cancer in which 1,008 heavy smokers and 1,074 nonsmokers in a certain city were followed for a number of years. The incidence of lung cancer over the total time period of the study among people exposed to the risk factor (cigarette smoking) is 183/1,008, or approximately 0.18 (18%), while the incidence among those not exposed is 12/1,074, or approximately 0.01 (1%).

TABLE 8-1

Risk Factor	Disease Outcome		
	Lung cancer (cases)	No lung cancer (controls)	Total
Exposed (heavy smokers)	183	825	1,008
Nonexposed (nonsmokers)	12	1,062	1,074
Total	195	1,887	2,082

The relative risk is therefore 0.18/0.01, or 18, showing that people who smoked cigarettes heavily were 18 times more likely to contract lung cancer than were nonsmokers. (Note that this is not a measure of *absolute* risk—it does not tell us the likelihood of heavy smokers contracting cancer in absolute terms.)

Because relative risk is a ratio of risks, it is sometimes called the **risk ratio**, or **morbidity ratio**. In the case of outcomes involving death, rather than just disease, it may also be called the **mortality ratio**.

Reports of risk reductions due to treatments in many clinical trials, and in almost all pharmaceutical advertisements, are of **relative risk reductions**; relative risk reduction (**RRR**) is equal to *1 minus relative risk*:

$$\text{Relative risk reduction} = 1 - \text{relative risk}$$

Relative risk reduction figures may be misleading if not understood properly. This can be illustrated by the well-known West of Scotland Coronary Prevention Study (WOSCOPS) (Shepherd et al., 1995):

> WOSCOPS was a double-blind randomized clinical trial, in which over 6,000 men with elevated cholesterol were randomly assigned to take either a placebo or the cholesterol-lowering drug pravastatin for an average of 4.9 years.

> There were 73 deaths from cardiovascular causes in the placebo group (3293 men); the cardiovascular mortality rate was therefore 73/3,293 = 0.022 (2.2%) in this group. In the pravastatin group (3,302 men), there were 50 cardiovascular deaths, giving a mortality rate of 50/3,302 = 0.015, or 1.5%. The relative risk of death in those given the drug is 1.5/2.2 = 0.68, so the relative risk reduction is (1 − 0.68) = 0.32, or 32%—showing that an impressive 32% of cardiovascular deaths were prevented by the drug.

> However, the **absolute risk reduction (ARR)** is the difference in absolute risk rates between the placebo and the drug, that is, 2.2% − 1.5% = 0.7%—a far less impressive-sounding figure, showing that of all men given the drug for 4.9 years, 0.7% of them were saved from a cardiovascular death.

Absolute risk reduction allows calculation of another statistic of clinical importance: the **number needed to treat (NNT)**. If 0.7% of patients' lives were saved by the drug, this implies that (100/0.7) = 143 patients would need to be treated to save 1 life. The formula for NNT is therefore:

Number needed to treat (NNT) = 100 / absolute risk reduction (ARR)

Looking at NNTs may allow the effectiveness of different treatments to be compared. The NNT allows a further calculation—the cost of saving one life with the treatment. WOSCOPS showed that 143 men needed to be treated for 4.9 years (58 months) to save one life. At the time the study was done, the drug cost approximately $100 per month in the United States, so it would have cost $100 × 58 = $5,800 to treat one man for this length of time in medication costs alone. It would therefore cost $5,800 × 143 = $829,400 to prevent one cardiovascular death over this period. By 2011, however, the drug cost less than $4 a month, so it would have cost $4 × 58 × 143 = $33,176 to prevent one death. These figures give a very different perspective on the value and cost-effectiveness of a treatment than simply saying that it reduces the risk of death by 32%. Similar analyses can be performed easily for almost any clinical trial in the literature.

Related to NNT is the concept of the **number needed to harm** (NNH). For example, in WOSCOPS, elevations in creatine kinase (CK) greater than 10 times the upper limit of normal were seen in 4 of the 3,302 men who took the drug, but in only 1 of the 3,292 men who took the placebo. The absolute risk *increase* of elevated CK is therefore (4/3,302) − (1/3,293) = 0.00121 − 0.00030 = 0.00091; so 0.09% of patients experienced this adverse effect as a result of being given the drug. This implies that 100/0.09 = 1,111 patients would need to be given the drug for one to be "harmed" in this way. NNH can be compared with NNT to evaluate the risks versus benefits of a given treatment.

ATTRIBUTABLE RISK

The attributable risk is the *additional* incidence of a disease that is attributable to the risk factor in question. It is equal to the incidence of the disease in exposed persons minus the incidence of the disease in nonexposed persons.

> In the previous example of lung cancer and smokers, the attributable risk is 0.18 − 0.01, or 0.17 (17%)—in other words, of the 18% incidence of lung cancer among the smokers in this study, 17% is attributable to smoking. The other 1% is the "background" incidence of the disease—its incidence among those not exposed to this particular risk factor.

Attributable risk is sometimes called **risk difference** or **cumulative incidence difference**, because it is the difference in the risks or cumulative incidences of the disease between the two groups of people.

From this kind of data, we can calculate what proportion of the overall population's burden of a disease is attributable to the risk factor. This is called the **population attributable risk (PAR)**, and it is equal to the *attributable risk multiplied by the prevalence of the exposure.*

> For example, if the risk of lung cancer that is attributable to smoking (its attributable risk) is 0.17, and the prevalence of exposure to smoking (i.e., the proportion of the population who have smoked heavily) is 0.4, then the PAR is 0.17 × 0.4 = 0.068; in other words, in the population as a whole, 7.2% of people have lung cancer that is attributable to smoking.

PAR therefore gives us a prediction of how much of the overall burden of disease in the population could theoretically be reduced if the risk factor were removed; in this case, if nobody in the population smoked, 6.8% of the total population would be spared from a lung cancer they would otherwise have developed.

If the cohort of smokers and nonsmokers were representative of the overall population in the city in which they live, and if the city had a population of 100,000, then eliminating smoking would theoretically prevent 6.8% of these people, or 6,800 people, from getting lung cancer; this is the **number prevented in the population**, also known as the **number of events prevented in a population**, or **NEPP**.

ODDS RATIO

Relative risk and attributable risk both require the use of cohort (prospective) studies, as shown previously. As noted in Chapter 5, cohort studies are generally expensive and time-consuming, and are therefore often impractical.

A common alternative to a cohort study is therefore a case-control (retrospective) study, in which people with the disease (cases) are compared with otherwise similar people without the disease (controls) (reviewed in Chapter 5). If the proportion of people who had been exposed to the possible risk factor is greater among the cases than the controls, then the risk factor is implicated as a cause of the disease.

Because the proportion of people in a case-control study who have the disease is determined by the researcher's choice, and not by the actual proportion in the population, such studies cannot determine the incidence or prevalence of a disease, so they cannot determine the risk of contracting a disease. However, the strength of the association between a risk factor and the disease can still be evaluated with a case-control study, by examining the proportions of the people with and without the disease who had, and had not, been exposed to the risk factor.

This is done through the use of the **odds ratio (OR)**. **Odds** is *the probability of an event occurring vs. the probability of it not occurring*, in other words $p / (1 - p)$.

> For example, the probability p of tossing a fair coin and getting heads is .5; the probability of not getting heads $(1 - p)$ is also .5, so the odds of getting heads is 1—you are as likely to get heads as not to get heads.

> Note that while probabilities can range only from 0 to 1, odds can range from 0 to infinity: the probability of tossing a coin and getting a head *or* a tail is essentially 1, and the probability of *not* getting a head or a tail is virtually 0, so the odds is 1 / 0, or infinity—you are infinitely more likely to get a head or a tail than not to get either!

Let us reexamine the data in Table 8-1, assuming now that this data was generated by a case-control study in which history of prior exposure to the risk factor (heavy cigarette smoking) was compared in 347 cases (with lung cancer) and 1,735 controls (without lung cancer).

> Of the 195 cases, 183 had been exposed to the risk factor; 12 had not been exposed. The probability (p) that a case had been exposed to the risk factor is therefore 183 / 195 = .94, and the probability that a case had *not* been exposed is 12 / 195 = .06 (i.e., $1 - p$).

> The odds that a case had been exposed is therefore 0.94 / 0.06 = 15.7. In other words, a person with lung cancer was 15.7 times as likely to have been a heavy smoker than to have not been a heavy smoker (numbers have been rounded for simplicity).

> Similarly, among the 1,887 controls, 825 people had been exposed to the risk factor, and 1,062 had not been; the probability that a control had been exposed is therefore 825 / 1,887 = .45. The probability that a control had *not* been exposed is 1062 / 1887 = .56; so the odds that a control was exposed is 0.45 / 0.56 = 0.81: a person without lung cancer was 0.81 times as likely to have been a heavy smoker than to have not been a heavy smoker.

The **odds ratio** (or **relative odds**) is the ratio of these two odds—it is the ratio of the odds that a case had been exposed to the odds that a control had been exposed:

$$\text{Odds ratio} = \frac{\text{odds that a case had been exposed to the risk factor}}{\text{odds that a control had been exposed to the risk factor}}$$

> In this example, the odds ratio is therefore 15.7 / 0.81 = 19.4. This means that among the people in this case-control study, a person with lung cancer was 19.4 times more likely to have been exposed to the risk factor (cigarette smoking) than was a person without lung cancer.

An odds ratio of 1 indicates that a person with the disease is no more likely to have been exposed to the risk factor than is a person without the disease, suggesting that the risk factor is not related to the disease. An odds ratio of less than 1 indicates that a person with the disease is *less* likely to have been exposed to the risk factor, implying that the risk factor may actually be a *protective* factor against the disease.

The odds ratio is similar to the relative risk: both figures demonstrate the strength of the association between the risk factor and the disease, albeit in different ways. As a result of their similarities, the odds ratio is sometimes called **estimated relative risk**—it provides a reasonably good estimate of relative risk *provided* that the incidence of the disease is low (which is often true of chronic diseases), and that the cases and controls examined in the study are representative of people with and without the disease in the population.

PREVENTIVE MEDICINE

All kinds of measures of risk are invaluable for informing policymakers' decisions about how to allocate resources for the benefit of a population's overall health, especially with regard to choices regarding **primary, secondary,** or **tertiary prevention**.

Primary prevention is pure prevention; it addresses the risk factors for disease *before* the disease has occurred. It may involve educating people about lifestyle changes (exercise, diet, avoiding tobacco use), reducing environmental factors (exposure to pollutants, toxins, etc.), immunization, or prophylactic medication. It is sometimes believed to be the most cost-effective form of health care, as it reduces the actual incidence of disease.

Secondary prevention involves identifying and treating people who are at *high risk* of developing a disease, or who already have the disease to some degree. Focused screening programs are of this kind: for example, ultrasounds to check for abdominal aortic aneurysms in male smokers, chest CT scans to screen for lung cancer in heavy smokers.

Tertiary prevention involves prevention of further problems (symptoms, complications, disability, death), or minimizing negative effects or restoring function in people who have *established disease*; examples include screening patients with diabetes for nephropathy, neuropathy, and retinopathy, and treating them to prevent these complications; or performing angioplasties or coronary bypass surgery on patients with known coronary disease.

EPIDEMIOLOGY AND OUTBREAKS OF DISEASE

Like epidemics, **outbreaks** are occurrences of a number of cases of a disease that are in excess of normal expectations—but unlike epidemics, outbreaks are limited to a particular geographic area. Epidemiological measures and methods contribute greatly to the understanding, identification, and treatment of outbreaks of disease.

Many diseases that have the potential to cause outbreaks are **reportable diseases**, which physicians in the United States are required to report to their state health departments, which in turn report them to the CDC. The vast majority of these are infectious diseases, such as anthrax, botulism, cholera, dengue fever, gonorrhea, hepatitis, HIV, Lyme disease, polio, rabies, syphilis, and so on; the only noninfectious nationally reportable diseases are cancers, elevated blood lead levels, food- and water-borne disease outbreaks, acute pesticide-related illnesses, and silicosis.

The national US 2012 reportable disease list (note that individual states' lists may differ) is available at http://wwwn.cdc.gov/nndss/document/2012_Case%20Definitions.pdf#NonInfectiousCondition.

Apart from reportable diseases, national and state agencies conduct ongoing surveillance of many other diseases and associated phenomena, including occupational diseases and injuries, over-the-counter medication use, illicit substance use, emergency room visits, and so on.

Although some outbreaks are identified through analysis of surveillance data, they are quite commonly identified as a result of direct reports by clinicians. Initially, a **cluster** of cases may be reported: this is a group of apparently similar cases, close to each other geographically or chronologically, which may be *suspected* to be greater than expected. It may not be clear if this is *actually* in excess of normal expectations, or even if the cases are actually of the same illness, so the actual existence of an outbreak may not always be clear initially.

For reportable diseases, local, state, or national records provide a baseline from which to judge if an outbreak has occurred, but for other diseases, it may be hard to tell if a given cluster is greater than the background incidence of a disease. This is particularly true if local or national publicity brings symptoms of a disease to public attention, resulting in increased presentation by patients and increased testing and reporting by physicians. The use of new diagnostic tests, or changes in the constitution of the population in question, may also account for an apparent "outbreak."

OUTBREAK INVESTIGATIONS

The first stage in modern outbreak investigations is to develop a **case definition**. In some situations (such as those of reportable diseases), the case definition is already well defined (and is listed in the CDC document mentioned above); in other cases, it may be very broad, and may be refined as the investigation progresses; it may be limited to particular people, places, and times, as well as to particular signs and symptoms (for example, all passengers and crew on a specific sailing of a specific cruise ship who developed vomiting and diarrhea, or all people with a specific lab finding).

On the basis of this, cases are **ascertained**, and further cases may be sought (e.g., by interviewing everybody who was on that ship at that time; or by interviewing, examining, or testing their friends and family members, or people who were on previous sailings of that ship).

With this data, outbreak investigations then describe the *times*, *places*, and *persons* involved, in order to rapidly identify the cause and transmission of the disease, with the immediate goal of controlling and eliminating it. In the longer term, these investigations may reveal new etiologic agents and diseases, or identify known diseases in areas where they did not previously occur, leading to methods of preventing future outbreaks. By identifying who contracts the disease and when and where it occurs, the disease may be brought under control, even if the actual etiology of the disease is not understood.

Times

To describe the chronology of an outbreak, an **epidemic curve** (or "**epi curve**") is created: This is actually a histogram showing the number of cases of the disease at each point in time, such as on a daily basis (shorter periods may be chosen if the disease appears to have a very short incubation period). The overall shape of the "epi curve" gives clues as to the nature of the outbreak.

A "curve" starting with a steep upslope, which then gradually tails off, suggests that the cases were all exposed to the same cause at around the same time: this would be a **common source epidemic**, such as from eating contaminated food, and the time between exposure and the midline of the "curve" would correspond to the incubation period of the pathogen.

- If the "curve" has a narrow, sharp peak with a rapid decline (Fig. 8-4), this suggests a **common** *point* **source epidemic**, such as a group of people at a banquet eating the same contaminated

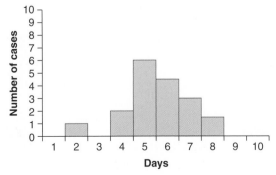

● **Figure 8-4** Epidemic curve suggestive of a common point source epidemic.

food at the same time. If the pathogen is known (e.g., through lab tests), the known incubation period can be used to infer the likely time of exposure.

• If the peak of the "curve" is rather broad (Fig. 8-5), this suggests a **common** *continuous* (or *persistent*) **source epidemic** in which the exposure lasted longer, such as where a contaminated product remained in the food supply chain for some time.

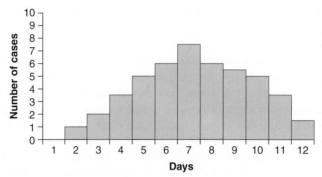

● **Figure 8-5** Epidemic curve suggestive of a common continuous (or persistent) source epidemic.

An irregular "curve" (Fig. 8-6) suggests a **common intermittent source epidemic**, in which people are exposed to the cause intermittently over a period of time.

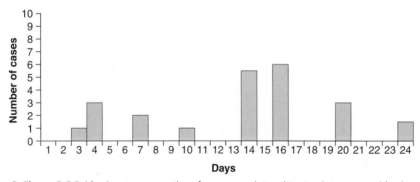

● **Figure 8-6** Epidemic curve suggestive of a common intermittent point source epidemic.

A "curve" with a series of progressively higher peaks (Fig. 8-7) suggests a **propagated epidemic**, in which each group of people who contract the disease then pass it on to another group.

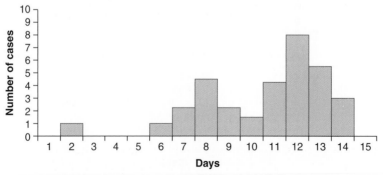

● **Figure 8-7** Epidemic curve suggestive of a propagated epidemic.

Places

A simple map of the cases of the disease (a "**spot map**") can show the relationship of cases to each other, and to potential infectious, toxic, or other environmental causes, whether natural or man-made. The map may be of any scale, as appropriate: a large geographical map, a local street map, or even a map of a small area such as an office building, a hospital, an aircraft cabin, or a ship.

> The classic example is the work of John Snow, an anesthesiologist in London, who marked the location of each case of cholera on a street map during an outbreak of the disease in 1854. He found that cases clustered around one particular public water source, the Broad Street pump, and also that two groups of people who lived near the pump were largely unaffected—they turned out to be brewery workers who drank alcohol rather than water (and what little water they drank came from a deep well at their workplace), and residents of a workhouse that had its own well. On the basis of this finding, the handle was removed from the Broad Street pump, and the epidemic was brought under control—decades even before the use of microscopes, let alone identification of the cholera bacterium and treatment with antibiotics.

Persons

Basic information about the people affected by the outbreak provides very important clues: age, gender, race, comorbidities, medication use, travel history, vaccination history, use of alcohol or drugs, what they ate and when, employment, recreation, sexual activity, substance use or abuse, what other people they are habitually in contact with (e.g., children, patients, immigrants), and so on.

Note that a person who has not been immunized against a particular disease may still be protected from it through the phenomenon of **herd immunity**: If a sufficient number of other people are immunized, then it is unlikely that the unimmunized person will come into contact with someone with the disease, as there will be only a small reservoir of people who potentially could have the disease. Related to this is the concept of **cocooning**: immunizing people who are likely to come into contact with a susceptible person, so that the susceptible person is unlikely to be exposed to the pathogen. For example, it is commonly advised that adults get immunized against pertussis if they are likely to come into contact with newborns, who are vulnerable until they have received at least some of their pertussis vaccines starting at the age of 2 months.

A measure that helps describe who gets sick in an outbreak of a disease is the **attack rate**: this is risk of contracting the disease; and it is actually the same as incidence, except that it is reported as the proportion of people who contract the disease in the specific outbreak in question. The attack rate is the *ratio of the number of people contracting a particular disease to the total number of people at risk*, expressed as a percentage:

$$\text{Attack rate} = \frac{\text{number of people contracting the disease}}{\text{total number of people at risk in the outbreak}} \times 100$$

For example, if 1,000 people ate at a barbecue, and 300 of these people become sick, the attack rate is $(300/1{,}000) \times 100 = 30\%$.

Calculating attack rates for different groups of people can help the source of an epidemic to be deduced.

> For example, Table 8-2 shows the attack rates for each food or combination of foods eaten by different people at the barbecue. The putative source of the illness can be deduced by inspecting the table for the maximum difference between any two attack rates. The largest difference between

TABLE 8-2

Food	Number Who Ate	Number Who Got Sick	Attack Rate (%)
Chicken only	100	25	25
Ribs only	80	10	12.5
Cole slaw only	20	7	35
Chicken and ribs only	200	18	9
Chicken and ribs and cole slaw	600	240	40
Total	1,000	300	30

any two attack rates is 31%: this is the difference between the lowest rate, 9% (in those who ate the chicken and ribs only), and the highest rate, 40% (in those who ate the chicken, ribs, and cole slaw). The implication is therefore that the cole slaw is the source.

Following the description of an outbreak in terms of the times, places, and people involved, a number of actions may be taken simultaneously:

- Measures to control the outbreak may be instituted, such as "engineering changes" (exemplified by removal of the Broad Street pump handle, designing nonreusable needles, etc.), recalling a food or drug from the supply chain, environmental controls, regulatory changes, or encouraging behavioral changes.
- Communication with those affected, or the public as a whole, about the outbreak.
- Laboratory testing, to identify the actual etiology (if it is not known) or to identify undiagnosed cases.
- Further case-finding, if others exposed to an apparent risk would benefit from diagnosis or treatment (e.g., trying to locate people who may have been exposed to a person with TB on a long-distance flight, or who may have received donated blood or organs from a person with a communicable disease).
- Development of hypotheses about the cause of the outbreak, and possibly planning further studies (such as case-control studies or cohort studies) to test these hypotheses. Note that entirely atypical outbreaks, such as the appearance of unusual pathogens in unusual locations, might raise the question of deliberate human causes, such as poisoning, sabotage, or bioterrorism.

Chapter 9

Ultra-High-Yield Review

Most USMLE Step 1 candidates probably spend only a very few hours reviewing biostatistics, epidemiology, and population health. A relatively short time should allow the student to memorize and self-test the ultra-high-yield items in the following checklist (Table 9-1), which is accompanied by mnemonics and reminders of memorable examples and other information to aid in recalling and using the material. Together with a background understanding from the rest of the book, these items should allow the student to pick up a good number of points in this increasingly important subject area.

TABLE 9-1

Understand/know the meaning of/be able to use	Example/mnemonic/notes	Page(s)
The four scales of measurement Addition and multiplication rules of probability Centiles Measures of central tendency: mean, mode, and median Confidence limits, including calculation of approximate 95% confidence limits Measures of variability: range, variance, and standard deviation	NOIR	
Proportions of the normal distribution that are within or beyond 1, 2, or 3 standard deviations from the mean	68, 95, 99.7%	
z scores		
Precision and accuracy	Dartboard	
Relationship between sample size and precision, how to increase precision and reduce the width of the confidence interval	Remember the square root sign	
How to be 95% confident about the true mean of a population Principles of hypothesis testing, establishing null and alternative hypotheses Meaning and limitations of p values and statistical significance		
Meaning of type I and type II errors in hypothesis testing and diagnostic testing How to avoid type I and type II errors in hypothesis testing	**I A's**—**A**lpha errors **A**ccept the **A**lternative **II BEAN**s—**B**eta **E**rrors **A**ccept the **N**ull	

(continues)

TABLE 9-1	(continued)	
Understand/know the meaning of/be able to use	**Example/mnemonic/notes**	**Page(s)**
Test power: how to increase it, and the dangers of a lack of it Differences between directional and nondirectional hypotheses; one- and two-tailed tests	Radar screen analogy	
Hazards of post hoc testing and subgroup analyses Effect modifiers and interactions	Aspirin effects and signs of the Zodiac	
Chi-square	Coin tossing; contingency tables	
Pearson correlation Correlation coefficients, r values, coefficient of determination (r^2)	Salt and blood pressure Avoid inferring causation!	
Spearman correlation Scattergrams of bivariate distributions	Birth order and class rank	
Simple linear regression	Using dye clearance to predict lidocaine clearance	
Multiple regression	Predicting risk of hepatic fibrosis in patients with fatty liver	
Logistic regression	Risk factors for the development of oropharyngeal cancer	
Survival analysis, life table analysis, the survival function, Kaplan-Meier analysis	4S trial: survival and simvastatin	
Cox regression and hazard ratios	Seven Countries study: smoking as a risk factor	

Choosing the appropriate basic test for a given research
  question
Different types of samples (simple random, stratified random,
  cluster, systematic) and problems of representativeness
	Memorize Table 4-1	

Problems of bias and lack of representativeness Phases of clinical trials Features of clinical trials, including control groups and blinding Handling missing ("censored") data: intention to treat, imputation and LOCF	1936 Presidential election opinion poll	
Noninferiority trials		
Descriptive or exploratory studies	Chimney sweeps, AIDS and Kaposi's sarcoma	
Advantages, disadvantages, and typical uses of: • cohort (incidence, prospective) studies • historical cohort studies • case-control studies • case series studies • prevalence surveys	Framingham study The mummy's curse DES and vaginal carcinoma Original 8 cases of AIDS and Kaposi sarcoma	

Understand/know the meaning of/be able to use	Example/mnemonic/notes	Page(s)
• ecological studies • postmarketing (Phase 4) studies The appropriate type of research study for a given question	COMMIT trial of quitting smoking	
The difference between POEMs and DOEs The principles of EBM (evidence based medicine) and the hierarchy of evidence The features of a systematic review, and the biases it minimizes Meta-analyses, funnel plots and forest plots Efficient literature searches for individual studies with the PICOS concept The difference between efficacy, effectiveness, and cost-effectiveness The factors that contribute to evidence of causality Meaning of validity (including internal and external validity), generalizability, and reliability	Torcetrapib: better lipids, worse outcomes Efficacy: Can it work?	
Sensitivity and specificity in clinical testing	"What epidemiologists want to know"; Table 7-6	
Positive and negative predictive values in clinical testing	"What patients want to know"; Table 7-6	
Accuracy in clinical testing What kind of test to use to rule in or rule out a disease How changing a test's cutoff point affects its sensitivity and specificity ROC curves and the area under the curve (AUC) The relationship between PPV, NPV, and prevalence	Table 7-6 "SNOUT" and "SPIN"	
Pretest probability, positive and negative likelihood ratios	Use of nomogram to determine posttest probability	
Use of prediction rules or decision rules	CHADS2 for risk of stroke with atrial fibrillation	
Decision analysis and decision trees	Medical vs. surgical management of carotid stenosis	
QALYs and cost-per-QALY Practice guidelines and their potential limitations	Prevention of hip fractures	
Population pyramids and their implications Mortality rates (including infant mortality and case-fatality rates)	Angola vs. Japan	
Adjustment or standardization of rates and the SMR	Florida vs. Alaska	
Epidemics and pandemics; endemic, and hyperendemic		
Relationships between incidence, prevalence, and mortality	The epidemiologist's bathtub	

(continues)

Understand/know the meaning of/be able to use	Example/mnemonic/notes	Page(s)
Absolute risk, relative risk, attributable risk, population attributable risk	Cohort study of smokers vs. nonsmokers with outcome of lung cancer	
Relative (RRR) and absolute (ARR) risk reduction, the difference between the two	WOSCOPS; misleading drug advertising	
Number needed to treat (NNT) and to harm (NNH)	WOSCOPS	
Odds ratios and their use in case-control studies	Case-control study of smokers vs. nonsmokers with outcome of lung cancer	
The differences between primary, secondary, and tertiary prevention		
The difference between disease clusters and outbreaks		
Principles of outbreak investigations, including case definition, case ascertainment, and the creation and interpretation of "epi curves" and spot maps	John Snow and the Broad Street pump	
Attack rates and their use in deducing the cause of an epidemic	Food poisoning at a barbecue	
Herd immunity and cocooning	Pertussis immunizations	

References

CHAPTER 3

ISIS-2 (Second International Study of Infarct Survival). *Lancet* 1988;2:349–360.

Lipid Research Clinics Program. The coronary primary prevention trial: design and implementation. *J Chronic Dis* 1979;32:609–631.

Sleight P. Debate: subgroup analyses in clinical trials: fun to look at - but don't believe them! *Curr Control Trials Cardiovasc Med* 2000;1:25–27.

CHAPTER 4

Angulo P, Hui JM, Marchesini G, et al. The NAFLD fibrosis score: a noninvasive system that identifies liver fibrosis in patients with NAFLD. *Hepatology* 2007;45:846–854.

D'Souza G, Aimee R, Kreimer AR, et al. Case-control study of human papillomavirus and oropharyngeal cancer. *N Engl J Med* 2007;356:1944–1956.

Jacobs DR, Adachi H, Mulder I, et al. Cigarette smoking and mortality risk: twenty-five year follow-up of the seven countries study. *Arch Intern Med* 1999;159:733–740.

Miettinen TA, Pyorala K, Olsson AG, et al. Cholesterol-lowering therapy in women and elderly patients with myocardial infarction or angina pectoris: findings from the Scandinavian Simvastatin Survival Study (4S). *Circulation* 1997;96:4211–4218.

Zito RA, Reid PR. Lidocaine kinetics predicted by indocyanine green clearance. *N Engl J Med* 1978;298:1160–1163.

CHAPTER 5

Fisher EB Jr. The results of the COMMIT trial: community intervention trial for smoking cessation. *Am J Public Health* 1995;85:159–160.

Herbst AL, Ulfelder H, Poskanzer DC. Adenocarcinoma of the vagina: association of maternal stilbestrol therapy with tumor appearance in young women. *N Engl J Med* 1971;284:875–881.

Nelson MR. The mummy's curse: historical cohort study. *BMJ* 2002;325:1482–1484.

Schulz KF, Altman DG, Moher D, et al. CONSORT 2010 Statement: updated guidelines for reporting parallel group randomised trials. *BMJ* 2010;340:c332.

CHAPTER 6

Barter PJ, Caulfield M, Eriksson M, et al. Effects of torcetrapib in patients at high risk for coronary events. *N Engl J Med* 2007;357:2109–2122.

Wanahita N, Chen J, Bangalore S, et al. The effect of statin therapy on ventricular tachyarrhythmias: a meta-analysis. *Am J Ther* 2012;19:16–23.

CHAPTER 7

Cummings SR, San Martin J, McClung MR, et al. Denosumab for prevention of fractures in post-menopausal women with osteoporosis. *N Engl J Med* 2009;361:756–765.

Fagan, TJ. Nomogram for Bayes's theorem. *New Engl J Med* 1975;293:257.

CHAPTER 8

U.S. Department of Health and Human Services, Centers for Disease Control and Prevention, National Center for Health Statistics, National Vital Statistics System. *Natl Vital Stat Rep* 2011;59(4).

Shepherd J, Cobbe SM, Ford I, et al. Prevention of coronary heart disease with pravastatin in men with hypercholesterolemia. *N Engl J Med* 1995;333:1301–1307.

Index

Note: Page numbers followed by f indicate figures; those followed by t indicate tables.

A

Abscissa, 5, 5*f*
Absolute risk, 92
Absolute risk reduction (ARR), 93
Accuracy, 20
Active post-marketing surveillance, 58
Addition rule, of probability, 2
Adjustment of rates, 89
Allocation bias, 48
Alpha error, 28
Alternative hypothesis
    definition of, 24
    directional, 31–32, 31*f*
    nondirectional, 31
Analytic studies, 53
Area of acceptance, 25
Area of rejection, 25
Ascertainment bias, 48
Assessor bias, 48
Attack rates, 99, 100*t*
Attributable risk, 94

B

Bar graph, 5*f*, 6
Beta error, 28
Between-subjects design, 52
Bias, 45
    precision and, 20, 20*f*
Bimodal distribution, 8, 8*f*
Binomial distribution, 2
Biomarkers, 59
Bivariate distribution, 36
Blocking randomization, 49

C

Case-control studies, 55–56
Case fatality rate, 89
Case report, 56
Case series studies, 56
Causality, 66
Censored observations, 40
Centile rank, 6–7
Central limit theorem, 16
Chi-square test, 34–35, 43
Citation bias, 61
Clinical trials
    control groups, 47–48
    definition of, 47
    effectiveness *vs.* efficacy, 65
    randomization, 48
Cluster, 96
Cluster samples, 46
Coefficient of determination, 38
Cohort studies, 53–55

Common intermittent source epidemic, 97
Common point source epidemic, 97–98
Common source epidemic, 97
Community intervention trials, 58
Community survey, 56
Confidence interval, 19
Confidence limits, 19
Confounders, 48
Confounding variable, 48
Continuous data, 3
Control groups, 47–48
Correlation
    definition of, 36
    negative, 36
    positive, 36
Correlational techniques, 36–44
Correlation coefficient
    definition of, 36
    determination of, 36–37
    types of, 37–38
Cost-effectiveness, 65
Cox proportional hazards analysis, 42
Cox regression, 42–43
Critical values, 25–26
Crossover designs, 52, 52*f*
Cross-sectional studies, 56
Cumulative frequency distribution, 4–5, 4*t*, 6, 6*f*, 7*f*
Cumulative incidence difference, 94
Cutoff point, 71

D

Deciles, 7
Decision analysis, 81–85
Decision criterion, 24–25
Decision rules, 80–81
Degrees of freedom (*df*), 22–23
Dependency ratio, 86
Dependent variables, 45
    causal relationship between, 66
Descriptive statistics, 1–14
Descriptive studies, 53
Detection bias, 48
Deviation scores, 10–11
Directional hypothesis, 31–32, 31*f*
Disability-adjusted life years (DALYs), 88
Discrete data, 3
Disease-Oriented Evidence (DOE), 59
Dissemination bias, 61
Distribution-free tests, 34
Dose-response relationship, 66
Double-blind studies, 48

E

Ecological study, 58
Effectiveness *vs.* efficacy, 65

Effect modifier, 33
Element, 1
Epidemic curve, 97
Epidemiology and population health
    disease frequency, measures of, 88
        adjustment of rates, 89
        epidemiologist's bathtub, 91–92
        incidence, 90
        mortality, 88–89
        prevalence, 90–91
    life expectancy, measures of, 88
    and outbreaks of disease, 96–97
    population pyramids, 86–87
    risk measurement
        absolute, 92
        attributable, 94
        odds ratio, 94–96
        preventive medicine, 96
        relative, 92–94
Errors (*see also* Standard error)
    type I, 28–29
    type II, 28–29
Estimated relative risk, 96
Estimated standard error, 21, 26
Evidence
    hierarchy of, 60, 60*t*
    meta-analysis, 61–63
    searching for, 63
    systematic reviews, 60–61
Evidence-based medicine (EBM), 60
Exclusion criteria, 60
Experimental groups, 47
Experimental hypothesis (*see* Alternative hypothesis)
Experimental studies, 46–51
    clinical trials, 47
    control groups, 47–48
    definition of, 46
    matching, 49
    randomization, 48
Exploratory studies, 53
Exposure allocation, 54
External validity, 50, 68

F

False-negative error, 28
False-positive error, 28
Follow-up studies, 54
Forest plots, 62, 63*f*
Frequency distributions, 3–8
    bimodal, 8, 8*f*
    cumulative, 4–5, 4*t*, 6, 6*f*, 7*f*
    definition of, 3
    graphical presentations of, 5–6, 5*f*
    grouped, 3, 4*t*, 5*f*
    J-shaped, 8, 8*f*
    relative, 3–4, 4*t*
    skewed, 8, 8*f*
Frequency polygon, 6, 7*f*
Funnel plot, 61, 62*f*

G

Gaussian distribution, 7, 7*f*
Generalizability, 50, 68
Gold standard, 68
GRADE system, 84
Gray literature, 60
Grouped frequency distribution, 3, 4*t*, 5*f*

H

H_0 (null hypothesis), 24
H_A (*see* Alternative hypothesis)

Hazard rate, 42
Health-adjusted life expectancy (HALE), 88
Herd immunity, 99
Heterogeneity, tests of, 62
Hierarchy of evidence, 60, 60*t*
Histograms, 5, 5*f*
Historical cohort studies, 55
Hypothesis
    alternative, 24, 31–32, 31*f*
    null, 24
    post hoc testing, 33–34
    subgroup analysis, 33
    testing of, 24–35

I

Inception cohorts, 54
Incidence, 90
Inclusion criteria, 60
Independent variables, 45
    causal relationship between, 66
Individual studies, searching for, 63–64
Infant mortality, 89
Inferential statistics, 1, 15–23
Information resources, patients referring to, 66–67
Informed consent, 51
Institutional Review Board (IRB), 51
Interaction, concept of, 34
Interim analysis, 52
Internal validity, 50, 68
Interval scale data
    definition of, 3
    statistical technique for, 43–44
Intervention studies (*see* Experimental studies)

J

J-shaped distribution, 8, 8*f*

K

Kaplan–Meier analysis, 41, 41*f*

L

Last observation carried forward (LOCF) method, 49
Life expectancy, 88
Life table analysis, 40–41
Likelihood ratios (LRs), 77–80
    negative, 78
    positive, 78
Linear relationship, 37
Logistic function, 39
Logistic regression, 39–40
Log rank test, 42
Longitudinal studies, 54

M

Mantel–Haenszel test, 42
Matching, 49
Mean, 9, 9*f*
    definition of, 9
    estimating standard error of, 21, 26
    population, 19–24
    probability of drawing samples with, 17
Mean square, 11
Measures of central tendency, 8–9, 9*f*
Measures of disease frequency, 88–92
Measures of effect, 92
Measures of life expectancy, 88

Measures of variability, 9–12
    range, 10
    variance, 10–11, 11*f*
Median, 8–9, 9*f*
Medical Subject Headings (MeSH), 64
MEDLINE database, 63
Meta-analysis, 61–63
Mode, 8, 9*f*
Morbidity, 91
Morbidity ratio, 93
Mortality, 88–89
Mortality ratio, 93
Multiple publication bias, 61
Multiple regression, 39
Multiplication rule, of probability, 2

N

Negative correlation, 36
Negative likelihood ratio (LR−), 78
Negative predictive value (NPV), 76–77, 77*t*, 78*t*
Negative studies, 61
Nominal scale data
    definition of, 3
    statistical technique for, 43
Nondirectional hypothesis, 31
Nonexperimental studies, 53–58
    analytic, 53
    definition of, 46
    descriptive, 53
    designs of
        case-control, 55–56
        case series, 56
        cohort, 53–55
        prevalence survey, 56
Non-inferiority trials, 52
Nonlinear relationship, 37, 38*f*
Nonparametric tests, 34
Nonrepresentative sample, 46
Normal distribution, 7, 7*f*, 12*f*
Normal range, 69
No-treatment control group, 47
Null hypothesis, 24
Number needed to harm (NNH), 94
Number needed to treat (NNT), 93

O

Observational studies (*see* Nonexperimental studies)
Odds ratio, 93*t*, 94–96
Old age dependency ratio, 86
One-tailed statistical test, 32
Ordinal scale data
    definition of, 3
    statistical technique for, 43
Ordinate, 5, 5*f*
Outbreaks, 96

P

p (*see* Probability)
Parametric tests, 34
Partially controlled clinical trials, 48
Participation bias, 45
Passive post-marketing surveillance, 58
Patient care, application to, 65
Patient-Oriented Evidence that Matters (POEMs), 59
Patient preference arm, 53
Pearson product-moment correlation, 37, 43
Pharmacovigilance studies, 57
Placebo control group, 48
Plausibility, 66
Population, 1, 15

Population attributable risk, 94
Population parameters, 15
Population pyramids, 86, 87*f*
Positive correlation, 36
Positive likelihood ratio (LR+), 78
Positive predictive value (PPV), 75
Positive studies, 61
Post hoc testing, 33–34
Post-marketing surveillance (PMS), 58
Post-test probability, 78, 79*f*, 80*t*
Power of statistical tests, 29–31
Precision, 19–21, 20*f*
Prediction rules, 80–81
Predictive techniques, 36–44
Predictive values
    definition of, 75
    negative, 76–77, 77*t*, 78*t*
    positive, 75
Prestratification randomization, 49
Pretest probability, 77
Prevalence, 90–91
Prevalence ratio, 56
Prevalence survey, 56
Preventive medicine, 96
Primary prevention, 96
Prior probability, 77
Probability
    definition of, 1
    of drawing samples with a given mean, 17
    post-test, 78
    pretest, 77
    samples, 1–2, 45
    z score for specifying, 14
Propagated epidemic, 98
Publication bias, 61, 62*f*

Q

Quality Adjusted Life Years (QALYs), 83–84
Quantiles, 7
Quartiles, 7

R

r^2 (*see* Coefficient of determination)
Randomization
    definition of, 48
    stratified, 49
Randomized clinical trials (RCTs), 48
Randomized controlled clinical trials (RCCTs), 48
Random samples
    cluster, 46
    simple, 46
    stratified, 46
    systematic, 46
Random sampling distribution of means, 15–17, 16*f*
Range, 10
Ratio scale data
    definition of, 3
    statistical technique for, 43–44
Receiver operating characteristic (ROC) curve, 74
Recruitment and retention diagrams, 49
Reference interval, 69
Reference range, 69
Reference values, 69–70
Referral bias, 45
Regression
    definition of, 38
    logistic, 39–40
    multiple, 39
    simple linear, 38
Regression coefficient, 39
Regression equation, 38–39
Regression line, 38

Relative frequency distribution, 3–4, 4t
Relative risk (RR), 61, 92–94
Relative risk reduction (RRR), 93
Reliability, 69
Repeated measures study, 52
Reportable diseases, 96–97
Reporting bias, 61
Representative sample, 45
Research ethics and safety, 51–53
Response bias, 48
Restriction, 49–51
Retrospective studies (*see* Case-control studies)
Reverse causation, 66
Risk
    absolute, 92
    attributable, 94
    relative, 92–94
Risk difference, 94
Risk ratio, 61, 93

S

Same-subjects design, 52
Sample, 15
Sample mean, 17–18
Sample statistics, 15
Sampling bias, 45
Sampling error, 15
Scattergram, 36–37, 37f
SD (*see* Standard deviation)
Secondary prevention, 96
Selection bias, 45
Self-selection, 45
Sensitivity, 70–71, 70t
Sensitivity analysis, 62
Significance level, 25
Simple linear regression, 38
Simple random samples, 46
Single-blind studies, 48
Skewed distributions, 8, 8f
Spearman rank-order correlation, 37, 43
Specificity, 71–74, 73f
Spot map, 99
Standard deviation, 11–12, 11f
    calculation of, 26
Standard error
    definition of, 17, 21
    determination of, 17
    estimation of, 21
    use of, 17–18
Standardization of rates, 89
Standardized mortality ratio, 89
Statistical significance, 28
Statistical tests
    one-tailed, 32
    power of, 29–31
    selection of, 43–44, 43t
    two-tailed, 31
Statistics
    descriptive, 1–14
    inferential, 1
    sample, 15
Stratification, 33–34
Stratified random samples, 46
Student's t, 22
Studies
    double-blind, 48
    experimental (*see* Experimental studies)

    negative, 61
    nonexperimental (observational) (*see* Nonexperimental studies)
    positive, 61
    searching for individual, 63–64
    single-blind, 48
Subgroup analysis, 33
Surrogate outcomes, 59
Survival analysis, 40–43
Survival function, 41
Systematic reviews, 60–61
    meta-analysis, 61–63
Systematic samples, 46

T

Temporal relationship, 66
Tertiary prevention, 96
Test-retest reliability, 69
Tests, of heterogeneity, 62
Time lag bias, 61
t scores, 21–22, 28
t tables, 22–23, 23t
t tests, 21
    for difference between groups, 32
Two-tailed statistical tests, 31
Type I error, 28–29
Type II error, 28–29
Types of data, 2–3

V

Validity
    definition of, 68
    external, 68
    internal, 68
Variability
    definition of, 10
    measures of, 9–12
Variables
    causal relationship between, 66
    confounding, 48
    definition of, 45
    dependent, 45
    independent, 45
Variance, 10–11, 11f
Volunteer bias, 45

W

Wait list control group, 52
Washout period, 52, 52f
Within-subjects design, 52

Y

Years of life lost (YLL), 88
Years of potential life lost (YPLL), 88
Youth dependency ratio, 86

Z

z score, 12–14, 13t, 14f, 21–22
z-test, 28, 43